T0180672

Electrothermal Frequency References in Standard CMOS

ANALOG CIRCUITS AND SIGNAL PROCESSING

Series Editors:
Mohammed Ismail, The Ohio State University
Mohamad Sawan, Polytechnique Montreal

For further volumes:
http://www.springer.com/series/7381

S. Mahdi Kashmiri • Kofi A.A. Makinwa

Electrothermal Frequency References in Standard CMOS

 Springer

S. Mahdi Kashmiri
Texas Instruments, Inc.
Delftechpark 19
Delft, The Netherlands

Kofi A.A. Makinwa
Delft University of Technology
Delft, The Netherlands

ISBN 978-1-4899-9525-4   ISBN 978-1-4614-6473-0 (eBook)
DOI 10.1007/978-1-4614-6473-0
Springer New York Heidelberg Dordrecht London

Printed on acid-free paper

Springer is part of Springer Science+Business Media (www.springer.com)

Acknowledgment

The original manuscript of this book was written as a Ph.D. thesis at Delft University of Technology, where I spent about five fruitful years. This Ph.D. journey would not have been possible without the support of many people, to whom I would like to express my gratitude.

At the foremost, I thank my supervisor Kofi Makinwa. Throughout years, Kofi's belief, enthusiasm and presence was a solid supporting force, which helped me go on. I learnt from him to be as critical as I can, and how to write and to present such that others can actually understand! I am thankful to him for these lifelong skills. I also thank Han Huijsing for his support and encouragement. I would like to thank Michiel Pertijs for his invaluable support with the design of the first reference's temperature compensation.

I need to thank Omid Shoaei from Tehran University, who first raised the passion for analog circuit design in me. He told us, his students, about the glory of writing to the red journal and about being fired if the company misses time-to-market due to design mistakes. It all seemed so exciting!

Academic experience during the Ph.D. research is extremely valuable. I am thankful to Marcel Pelgrom and Lucien Breems for the opportunity of assisting them within their data converter design courses at TU Delft.

I would like to thank Greta Milczanowska and Marc van Eylen from Europractice, IMEC, for their support with the manufacture of 0.7 μm chips. I express my gratitude to Frank Thus, Paul Noten and Erik Moderegger for their invaluable support and NXP Semiconductors for fabrication of the scaled electro-thermal frequency references in 0.16 μm CMOS.

I am thankful to the wonderful room-mates at the electrical engineering faculty of TU Delft, with whom I shared lots of memories. In a chronological order: André Aita, Lukas Mol, Ferran Reverter, Caspar van Vroonhoven, Luca Giangrande, Saleh Heidary, and Zichao Tan. Caspar worked on thermal-diffusivity-based temperature sensors, which is why we enjoyed lots of fruitful discussions, technology and experience sharing, as well as a wonderful camping trip to New Zealand.

Special thanks to the staff of the Electronic Instrumentation (EI) lab of TU Delft, whose valuable work allows the department to run. During my times Inge, Trudie, Pia, Helly, Ilse and Joyce ran the secretary office. Thanks to Willem van der Sluys who made it financially feasible. Thanks to the technical staff Ger, Piet, Jeroen, Maureen, and Zu-Yao Chang. Thanks to Antoon Frehe who took care of the servers and made designing possible.

And of course, many friends and colleague whose presence in my life has meant a lot to me. In an alphabetical order: Arvin Emadi, Berenice, ChungKai Yang, Dafina Tanase, Eduardo Margallo, Frerik Witte, Gayathri, Gregory Pandraud, Kamran Souri, Lukasz, Martijn Snoeij, Morteza Alavi, Mohammad Talaie, Mohammad Mehrmohammadi, Mohammad Farazian, M. Nabavi, Nishant, Omid Noroozian, Pedram Khalili, Paulo Silva, Qinwen Fan, Rong Wu, Sha, Sharma, Shishir, Ugur, Wen Wu, Youngcheol Che, Yue Chen and Zili. Also thanks to the wonderful remote colleagues at NXP: Fabio Sebastiano and Mohammed Bolatkale.

Thanks to Sarah von Galambos for English corrections of the original Ph.D. thesis draft.

The preparation of the original thesis draft took one and half year during which I combined working and writing. This would not have been possible without the support of my colleagues at the precision systems group of Texas Instruments Delft design center (former National Semiconductor). I am especially thankful to my manager Wilko Kindt. I am also thankful to my colleagues: Frerik Witte, Jinju Wang, and Sergio Roche.

I am very much indebted to my family, my parents and my brothers, for their support and trust in me. It is thanks to their dedications and love that I have been able to pursue my dreams. Thanks to all of you.

And last but not the least: I am indefinitely thankful to my fiancée, Esmée. Without her unconditional love, understanding, and support I would not have finished this book.

Delft, The Netherlands   S. Mahdi Kashmiri

Contents

Chapter 1
Introduction

The operation and performance of electronic systems depends on the accuracy of timing signals. For decades, these signals have been generated by quartz crystal oscillators, which cannot be integrated into microchips. A lot of efforts have been dedicated to the realization of integrated frequency references, resulting in various types of silicon-based frequency references. Some combine a MEMS (Micro Electro Mechanical System) resonator with a silicon chip, while others rely on integrated resistors, capacitors, or inductors. This book describes an alternative approach to the realization of integrated frequency references. Unlike others, it does not rely on the accuracy of on-chip electrical components. Instead, it relies on the thermal properties of bulk silicon, harnessed by electrothermal structures. The operation of such structures is governed by a physical parameter: thermal-diffusivity, a measure of the rate at which heat diffuses through a silicon substrate. This book shows various standard CMOS prototypes with output frequencies up to 16 MHz and stabilities of $\pm 0.1\%$ over the military temperature range ($-55°C$ to $125°C$). Realizations in scaled processes show that electrothermal frequency references benefit from CMOS scaling.

This chapter is an introduction to the book. It provides a brief discussion of various frequency related parameters, an overview of quartz crystal oscillators and the efforts towards their replacement. Thermal oscillators are then introduced as the predecessors of electrothermal frequency references. Furthermore, the concept of thermal diffusivity and the way it can be harnessed in a standard CMOS process is described. The chapter ends by presenting the motivations, challenges and organization of the book.

1.1 Frequency and Its Accuracy Measures

Frequency and time are interdependent parameters. Everyone has a feeling of time, a parameter that has been measured historically by keeping track of natural phenomena such as the cycles of day, night, and seasons [1]. The unit of time is the second,

S.M. Kashmiri and K.A.A. Makinwa, *Electrothermal Frequency*
References in Standard CMOS, Analog Circuits and Signal Processing,
DOI 10.1007/978-1-4614-6473-0_1, © Springer Science+Business Media New York 2013

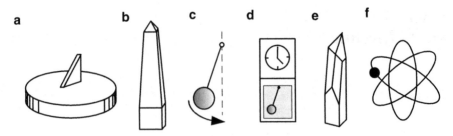

Fig. 1.1 (**a**) Sundials, (**b**) obelisks, (**c**) pendulum, (**d**) pendulum clock, (**e**) quartz crystal, and (**f**) Cesium atomic clock

the smallest quantity that most wrist-watches can indicate. Frequency, denoted by symbol f, is the number of occurrences of a periodic event within 1 s [2]. The unit of frequency is Hertz (Hz), meaning that a periodic event that occurs once per second has a frequency of 1 Hz. For instance, a violin string that produces an E musical note, vibrates 660 times in a second, corresponding to a frequency of 660 Hz. Furthermore, the duration of one cycle of a repetitive event is its period, which is denoted by symbol T. The period T is equal to the reciprocal of the frequency f.

In ancient times (thousands of years BC), the Egyptians used obelisks and sundials to keep track of time (Fig. 1.1) [3]. Later, it was discovered that an object with a reliable periodic movement (stable frequency) could also be used to keep time. One of these early objects was the pendulum (Fig. 1.1). In 1583, Galileo discovered that a pendulum swings with a nearly constant period. Later, in 1656, Huygens invented the pendulum clock. After the invention of quartz-crystal oscillators in 1918, it was possible to embed quartz crystals with oscillation frequencies of 32,768 Hz into electronic wrist watches [1]. This specific frequency can be easily related to a 1 Hz event via a binary counter that divides the oscillator frequency by 2^{15}. Cesium atomic clocks were developed in the 50s. This invention enabled extremely accurate time (frequency) measurements, with an inaccuracy of one second in ten million years [1]!

Most of us encounter the concept of frequency on a daily basis in the form of musical notes, radio frequencies, etc. However, these are not the only ways that frequency plays a role in our modern lives. In the current era of information, we send and receive data through our communication devices. It is because of the accurate and stable generation and detection of frequencies that we have mobile phones, wired and wireless data networks, ever faster computers, navigation systems based on the Global Positioning System (GPS), etc. In all these systems, accurate frequency sources allow data from different channels to be combined, sent through a communication medium, and successfully received at the destination.

The accuracy and stability of frequency references are crucial to the operation of any instrument that uses them. A faulty or broken tuning fork produces a resonant frequency that deviates from the desired musical tone. A music instrument tuned with that fork might then irritate a musician's sensitive ears. If the local oscillator in an FM radio has an unstable frequency that jumps around every now and then, the received signal will, accordingly, jump back and forth between the various

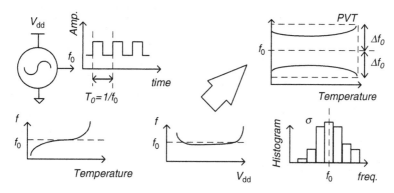

Fig. 1.2 Effects of process, voltage, and temperature variations on an oscillator's output frequency (PVT effects)

channels. If the same oscillator has a poor noise performance, then you might hear the news in the background of your desired jazz music. If the crystal oscillators in your cell phone are too temperature dependent, your call will drop every time you step out of your warm office onto the cold snowy streets. Finally, if the frequency reference in your MP3 player does not meet the USB standard, copying an MP3 file through the USB port of your laptop might require several attempts.

The abovementioned examples tell us that a frequency reference needs to have a certain level of accuracy, stability and noise performance [4]. This means that parameters should be defined to describe the quality of such a reference. In principle, an electrical frequency reference can be regarded as a circuit block that receives a supply voltage V_{DD} and produces a periodic output signal at a target frequency f_0 (Fig. 1.2). The stability of a frequency reference can be considered from various points of view. One such view involves observing how the output frequency f_0 varies as a function of environmental parameters such as ambient temperature and supply voltage V_{DD}. Furthermore, variations in the fabrication process can shift the value of f_0. The combined effect of these parameters (process, voltage, and temperature) is called the PVT tolerance of a reference.

The graphical illustrations in Fig. 1.2 show how each of the PVT elements can affect the nominal frequency of a frequency reference. Over its full range of variations, each parameter causes a certain amount of deviation in the nominal value of f_0. When the effects of all parameters are added, a total deviation of Δf_0 can be expected in the reference's frequency. This is a proportion of the nominal oscillation frequency f_0, and thus the ratio $\Delta f_0/f_0$ is normally denoted in percentage (%) or in parts-per-million (ppm). This is a measure of the frequency reference's stability and determines its overall level of inaccuracy.

Apart from the stability or accuracy of a frequency reference, its output also includes noise. This noise adds uncertainty to the period of oscillation. This is shown in Fig. 1.3, where a few cycles of a square-wave clock are illustrated. In this figure, the period of oscillation has a random variation due to noise, which usually has a Gaussian distribution [5–7]. The average value of this distribution has a mean,

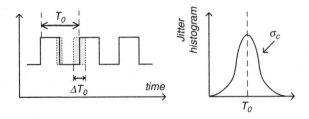

Fig. 1.3 Variations in the period of oscillation due to noise and the jitter histogram

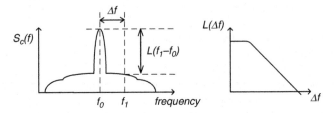

Fig. 1.4 Phase noise of an oscillator

which is the average period of oscillation called $T_0 = 1/f_0$. The standard deviation of this distribution, σ_C, is defined as the cycle-to-cycle jitter, which is a measure of the magnitude of period fluctuations. Jitter is normally defined as a root-mean-squared (rms) quantity.

Jitter is a time-domain means of quantifying the noise in an oscillation period, but this can also be done in the frequency domain. The associated parameter is called phase noise [5], and is especially important for sine-wave oscillators used in telecommunication applications [7]. Ideally, the frequency domain representation of a sine-wave signal has a power spectral density in the form of a peak occurring at the frequency of oscillation, f_0. Due to noise in the phase of this sinusoidal signal $S_C(f)$, its power spectrum will exhibit a "skirt" as shown in Fig. 1.4. The amplitude of this skirt at frequencies with a certain offset with respect to f_0 is an important parameter in the design of radio receivers. In such applications, the local oscillator's phase noise will affect the receiver's selectivity [8]. The phase noise spectrum $L(\Delta f)$, is defined as the attenuation in dB referenced to the value of $S_C(f)$ at f_0 and at an offset frequency $\Delta f = f_1 - f_0$. This is normalized to the main carrier's power and denoted in dBc/Hz (see Fig. 1.4).

1.2 Challenge of Integrating Frequency References

Historically, crystal oscillators have been the only frequency control components that could achieve high performance with regard to accuracy and noise [9]. However, due to the dedicated manufacturing process of a quartz crystal, they are nearly impossible to realize in IC technologies. For years, integration has brought more reliability to electronic systems allowing for lower prices and more functionality at smaller form factors. This is thanks to the reliable and large volume production made available by IC technologies such as the CMOS process, which have advanced with the evolution of microprocessors. To benefit from this integration trend, researchers and circuit designers have faced the challenge of producing frequencies as stable as those made by quartz crystal oscillators, but only through the use of on-chip circuitry [10].

Integrated (silicon-based) frequency references need to rely on the properties of on-chip elements in order to produce accurate time constants. Of the various methods that have been developed in the past decades, only a few have been commercialized so far. One of them is the silicon MEMS resonator-based oscillator, in which the quartz crystal is replaced by a MEMS resonating structure. The resonator is then attached to another silicon die with the circuitry that maintains the oscillation and performs the temperature compensation [11]. MEMS based frequency references with sub-ppm stabilities are now available commercially [12]. They can replace quartz crystal oscillators with packages that have exactly the same foot-print. The major difficulty with this technology is the special processing required for the fabrication of MEMS structures, which makes their integration with baseline IC technologies such as standard CMOS uneconomic. This usually leads to a solution requiring two dies within one package.

Standard CMOS compatibility of an integrated frequency reference makes it possible to combine it with larger systems-on-a-chip. Such complicated systems combine accurate analog functionalities with sophisticated digital signal processing in a single die and are normally implemented in CMOS process. Various standard-CMOS-compatible frequency references have been introduced so far, mainly relying on passive elements such as resistors, capacitors, and inductors. Among these, LC oscillators (using inductors and capacitors in a resonant circuit) [13] have been commercialized. These oscillators achieve accuracies in the order of a few hundred ppm by means of trimming and temperature compensation. RC oscillators dissipate less power than the LC oscillators; however, their accuracy is limited to tens of thousands of ppm [14]. Although, this certainly does not compete with the accuracy of quartz crystals, RC oscillators are often used in low-power applications such as biomedical implants.

Traditionally, the methods of realizing integrated frequency references make use of the generation, transfer, and processing of signals in the electrical or mechanical energy domains. RC and LC oscillators are purely electrical, while MEMS based resonators are electro-mechanical parts. The frequency generated by each of these methods depends on the manufacturing process and on the effect of temperature variations. These dependencies necessitate some means of trimming and temperature compensation in order to achieve reasonable accuracies. Sometimes, the lack of correlation among the various sources of variation requires multi-point temperature trimming, which increases the manufacturing costs.

Questions that we might ask ourselves could be: "Is there another physical property of silicon (besides the electrical-domain properties) that is stable enough and can be used as a means of producing time (frequency) references? Could this property be harnessed by means of electronic circuitry? Is this property something that can be used in any IC technology, especially within the standard CMOS process? Is it a property with reproducible behavior in response to environmental parameters such as temperature? Is a research plan for investigating the possibilities and limitations of using this property for the goal of on-chip frequency generation attractive?"

1.3   Frequency Generation Based on the Thermal Properties of Silicon

Energy can be produced, processed and transferred in any of the five physical domains. These include the electrical, mechanical, chemical, thermal and electromagnetic domains. Efforts to generate stable on-chip frequency references have, so far, mainly concentrated on the electrical, mechanical and the electromagnetic domains. The *thermal-domain* properties of silicon have received much less attention.

The first publications in this field date back to the 1970s, when the transport of thermal signals in microelectronic structures was investigated. In 1971, Gray and Hamilton showed [15] that the interactions between electrical and thermal-domain signals can be used to produce large time constants in silicon integrated circuits. Their main interest was to produce filters with very low cut-off frequencies. These efforts resulted in microstructures in which integrated heaters were fabricated close to integrated thermal sensors and within a silicon substrate (see Fig. 1.5). In such structures, the heat dissipated in the heater diffuses through the substrate and is sensed by the temperature sensor after a certain *thermal delay*. This delay is associated with the substrate's thermal inertia. Hamilton used this technique to realize integrated high-Q band-pass filters with bandwidths ranging from a few Hertz to a few hundred Hertz [16].

In 1972, Bosch, a researcher at Philips Research Laboratories, proposed [17] the use of the Seebeck (thermoelectric) effect, i.e. the direct conversion of temperature difference to electric voltage, to realize on-chip temperature sensors in the form of thermocouples [18]. These were made by "p" or "n" type semiconductor materials in contact with Aluminum. He considered locating a heater about 200 μm apart from a thermocouple [18]. An amplifier fed back the output of this thermocouple to the heater, forming a thermal oscillator. Bosh reported a nominal oscillation frequency of 200 kHz, but did not publish measurement results describing the effect of process and temperature variations. Further work on thermal oscillators was published in 1995 by Szekely [19], in which the behavior of a thermal relaxation oscillator was investigated over temperature. This was with the special interest of studying the possibility of using thermal oscillators as temperature-to-frequency converters.

The concept of thermal oscillators builds on the well-defined thermal delays that can be realized in microstructures. Such delays involve the transfer of heat within a

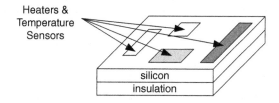

Fig. 1.5 Micro heaters and temperature sensors diffused into a silicon substrate

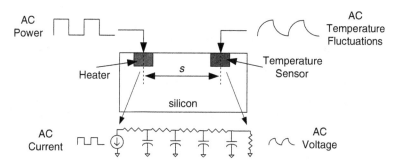

Fig. 1.6 Silicon slab with a heater and a temperature sensor implemented in it at a distance s and the electrical circuit equivalent to it

defined geometry, fabricated in a silicon substrate. The substrate acts as the heat transferring medium. Figure 1.6 illustrates the side view of a silicon slab, in which a diffusion heater, e.g. a resistor, is implemented in close proximity (s is a few tens of microns) to a relative temperature sensor, e.g. a thermopile. The heat generated in the heater diffuses through the substrate and results in a local temperature change. This is mediated by phonons [20] and thus involves mechanical vibration within the silicon atoms of the lattice.

The rate at which heat diffuses through the substrate is determined by the thermal-diffusivity of silicon, D (in cm^2s^{-1}) [21, 22], which is a temperature dependent parameter. This dependency is associated with the effect of temperature on the silicon lattice causing its expansion or contraction and affecting the mechanical vibration of the crystalline silicon atoms. In the past, the temperature dependence of D has been characterized over various temperature ranges. These are summarized in [22], indicating an approximate relation of $T^{-1.8}$ (where T is the absolute temperature) [22, 23]. This implies that the thermal delay resulting from the thermal diffusivity of silicon will also be a function of temperature.

If AC power is dissipated in the structure shown in Fig. 1.6, the resulting temperature fluctuations at the temperature sensor will be translated back into an AC electrical signal. However, the phase of this signal will be delayed compared to the heater's power. This delay is a function of D and s. Such a structure behaves like a low-pass filter and is therefore called an electrothermal filter (ETF). Like an electrical filter, an ETF has a defined phase versus frequency characteristic. In principle, such a structure can be implemented in any IC process.

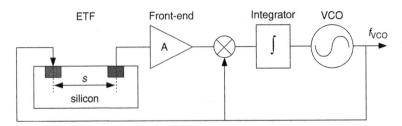

Fig. 1.7 Electrothermal frequency-locked loop (FLL) with an analog integrator

In 2006, Makinwa showed that by embedding an ETF in a frequency-locked loop (FLL), a voltage-controlled-oscillator (VCO) can be slaved to the thermal-diffusivity of silicon (see Fig. 1.7) [23]. The ETF's heater is driven by the VCO output. The ETF's output signal is then demodulated by the same heater drive signal via a synchronous demodulator. The output of the demodulator is integrated and used to drive the VCO. Feedback forces the VCO to oscillate at a frequency where the output of the demodulator is zero. This corresponds to an ETF phase shift of 90°. This means that ideally, the accuracy as well as the characteristics of the output frequency will be solely determined by the ETF.

This architecture represented a major breakthrough compared to the early thermal oscillators. This was mainly due to the use of synchronous demodulation and integration. One of the main drawbacks of the earlier thermal oscillators was their poor jitter performance. This is because silicon is a good conductor of heat and even at large heater power levels the thermopile signal is rather small. In the presence of the wideband thermal noise produced by the thermopile's resistance, the signal-to-noise ratio at the output of an ETF is quite poor. The narrow-band tracking filter employed in the electrothermal FLL of [23] reduces the noise bandwidth and achieves a low level of jitter. The work in [23] was initially aimed at developing a temperature-to-frequency converter. It demonstrated that the output frequency of an electrothermal FLL can be successfully locked to D, and thus exhibit the same temperature dependence [21, 22].

In [23] a remarkable level of untrimmed inaccuracy, i.e. a device-to-device output frequency spread of $\pm 0.25\%$ (3σ) was reported over the industrial temperature range. This was very promising because, to first order, this is determined by the ETF's phase accuracy, which is a function of D and its geometry. The accuracy of the geometry is defined by the photolithography used in the CMOS process, while the value of D should be stable for the doping levels used for the IC-grade silicon substrates [22, 23]. These results showed that perhaps the thermal-diffusivity of silicon could be a potential basis for on-chip frequency generation. However, the output frequency of the electrothermal FLL has the same temperature dependence as D. To build an integrated frequency reference, stability over temperature is required, and so a means of temperature compensation is necessary.

1.4 Motivation

The main *practical* motivation of the work described in this book is to make an integrated frequency reference with no external components, which can be fabricated in standard CMOS process. As described earlier in this chapter, the elimination of quartz crystal oscillators, as the last external electrical components, has motivated a large amount of research in the past years. Many of these efforts have resulted in silicon-based on-chip frequency references that either rely on the tolerance of on-chip passive elements, or require special manufacturing processes. However, only the MEMS-based and the LC-based oscillators have been commercialized.

The main *scientific* motivation of the work in this book is to explore the feasibility and the level of stability that can be achieved by an alternative method of on-chip frequency generation. Unlike the conventional methods, the proposed method does not rely on the accuracy of on-chip electrical elements. Instead, the thermal diffusivity of silicon is harnessed through standard CMOS compatible structures called electrothermal filters (ETF). The thermal diffusivity of silicon is defined by the parameter D, which is the rate at which heat diffuses through a silicon substrate. An ETF is a low-pass structure whose phase response is determined by D and by geometry. The application of ETFs in frequency-locked loops facilitates the realization of an electrical oscillator locked to D. The stability of the output frequency is then no more determined by the oscillator's own tolerances and drift, but determined by the ETF characteristics.

The work done on thermal-diffusivity-based (TD) temperature sensors has shown that ETF tolerances are in the order of 0.1% [23, 24]. This is mainly determined by the purity of the silicon substrate and by the accuracy of the lithography used in the IC technology. For scaled processes, with smaller feature sizes, the lithographic accuracy improves, which implies that ETFs should benefit from Moore's law [25]. This agrees with the observation that the untrimmed accuracy of TD temperature sensors improves as a function of CMOS scaling [24, 26]. Besides exploring the achievable levels of performance by electrothermal frequency references, another motivation for this work is to show that such references can also benefit from process scaling.

An electrothermal frequency reference needs an accurate means of temperature compensation. This is because of the temperature dependence of D. Contrary to the multi-point temperature trims used in some silicon-based frequency references, the temperature compensation of an electrothermal frequency reference should only require a single room-temperature trim to compensate for lithographic errors. This is crucial in reducing the extra costs associated with the test time.

The performance metrics of interest in this work are process and temperature spread and the achievable levels of output jitter. Since this method has not been explored yet, there are no specific performance targets for the, to be developed frequency references. From a purely scientific exploration point of view, design methodologies can be devised with the aim of discovering the limits of performance, which should then be confirmed by experimental results. This book thus describes a *pioneering* effort that aims to determine the possibilities and limitations of the proposed method. It represents the first steps in the evolutionary path of electrothermal frequency references.

1.5 Challenges

The design and implementation of an electrothermal frequency reference involves several challenges at both the system and circuit levels. These mainly involve accuracy and noise related tradeoffs. As mentioned earlier, the expected tolerance of an electrothermal filter (ETF) is about 0.1%, which should not be altered by its interface circuitry. This means that the frequency-locked loop (FLL) in which the ETF is embedded should be able to excite it and readout its output signal accurately. The phase information contained in this signal needs to be processed with accuracies in the order of tens of milli-degrees. Precision analog circuit design techniques then need to be applied in order to suppress extra error sources such as excess electrical phase shift and residual offsets.

The simplified FLL shown in Fig. 1.7 is based on the proposal in [23]. This loop involves an analog integrator that determines the loop's narrow noise-bandwidth and suppresses the ripple associated with the synchronous phase detector used to readout the ETF phase. To achieve this, an external 1 μF capacitor was used in [23]. The elimination of this external component is a major system-level challenge. To achieve this, a new digitally-assisted FLL is proposed in this work. The narrow bandwidth is achieved through a digital loop filter whose inclusion in the loop, however, requires the addition of analog-to-digital and digital-to-analog conversions.

Without temperature compensation, an electrothermal FLL exhibits a temperature dependence in the order of 3000 ppm/°C (at room temperature). To guarantee 0.1% frequency accuracy, a temperature compensation scheme with a state-of-the-art inaccuracy of about 0.1°C is required. This has, so far, been achieved only by band-gap temperature sensors, which are based on the temperature dependence of bipolar transistors [27]. Therefore, an on-chip band-gap temperature sensor combined with a delta sigma data converter was used to measure the temperature of the die. This could then be injected into the digitally-assisted FLL through a digital mapping scheme in order to compensate the frequency reference.

The design of an electrothermal frequency reference involves trade-offs between accuracy, output frequency and jitter performance. An ETF's geometry determines its thermal delay. The smaller the geometry, the smaller the delay and hence the higher the output frequency of the reference can be. On the other hand, the accuracy of an ETF's phase shift determines the accuracy of the output frequency. Since this is, to first order, determined by lithographic error, reducing the geometry increases its effect. This means that the ETF's dimensions have to be increased in order to improve its intrinsic accuracy. However, silicon is a good conductor of heat, and so increasing these dimensions reduces its output signal. The lower the signal, the more will be the effect of the ETF's wideband thermal noise on the FLL's jitter. Therefore, there are trade-offs among the accuracy, output frequency and jitter of an electrothermal frequency reference.

1.6   Organization of the Book

This book describes an alternative method of on-chip frequency generation based on the thermal diffusivity of silicon. Apart from this introductory chapter, the second chapter provides a literature study on state-of-the-art silicon-based frequency references. For each approach, a brief introduction to the history, principles of operation, state-of-the-art realizations, performance measures, and the associated possibilities and limitations will be provided. The study covers silicon MEMS resonator based oscillators, as well as LC, RC, relaxation, ring, and electron-mobility-based frequency references.

Chapter 3 provides an overview of the concept of on-chip frequency generation based on the thermal properties of silicon. The thermal-diffusivity constant of silicon, D, will be introduced. It will be shown how an electrothermal filter (ETF) can harness this physical property. ETFs in standard CMOS and their design parameters will be described. An overview of the earlier thermal oscillators and their limitations will be provided. Furthermore, an electrothermal frequency-locked loop (FLL) will be introduced as a system level solution to the drawbacks of early thermal oscillators. It will be shown why an FLL is a suitable foundation for building an electrothermal frequency reference. The dynamics of the FLL as well as the effect of the ETF thermal noise on its output jitter will be analyzed. Also, the earlier generations of CMOS FLLs and the challenges associated with their integration will be reviewed. This motivates the need for the realization of an alternative FLL.

Chapter 4 describes a new architecture for electrothermal FLLs, which is more suitable for CMOS integration. The proposed digitally-assisted FLL (DAFLL) achieves the required narrow noise-bandwidth by means of a digital loop filter. The proposed system-level architecture of the loop will be introduced. Furthermore, the design, implementation and characterization of the DAFLL will be covered. This includes a phase digitizer in the form of a phase-domain $\Delta\Sigma$ modulator (PD$\Delta\Sigma$M) and a digitally-controlled oscillator (DCO). The design and characterization of these blocks will be described in the framework of two test chips.

Chapter 5 describes the complete implementation of the first electrothermal frequency reference in a 0.7 μm standard CMOS process. This includes the addition of temperature compensation to the DAFLL described earlier with the help of an on-chip band-gap temperature sensor. As a result, the reference produces an output frequency of 1.6 MHz and is stable to ±0.1% over the military temperature range (−55°C to 125°C). The system-level considerations of the temperature compensation scheme are presented. Then the system and circuit level design of the band-gap temperature sensor will be described in detail. Finally, the characterization results on a test chip including the complete frequency reference will be provided.

Chapter 6 describes the design and implementation of a scaled electrothermal frequency reference in a 0.16 μm standard CMOS process. The aim of this implementation is to demonstrate the feasibility of electrothermal frequency references in modern CMOS process, as well as to demonstrate that such references can benefit from technology scaling. For a given accuracy, the improvements achieved by scaling include less jitter, greater output frequency, and less power consumption and chip area. The scaling strategy starting from the ETF and extending to the analog circuit design will be described. The system and circuit design as well as the experimental results on a test chip implemented in a 0.16 μm CMOS process will be provided. The scaled reference dissipates 2.1 mW from a 1.8 V supply (3.7× reduction compared to the previous generation), generates a 16 MHz output frequency (10× higher), and is stable to ±0.1% over the military temperature range. Its 45 ps rms period jitter is 7× lower and its area is 12× smaller than the previous implementation.

In Chap. 7, the main conclusions of the book are summarized. Furthermore, possible future work on electrothermal frequency references will be described.

References

1. Allan D et al (1997) The science of timekeeping. HP Application Note 1289
2. The Wikipedia page on frequency to be found on-line at: http://en.wikipedia.org/wiki/ Frequency
3. Jespersen J, Fitz-Randolph J (1999) From sundials to atomic clocks: understanding time and frequency. National Institute of Standards and Technology, Dover Publications, Inc., Mineola, New York
4. IEEE Standard definitions of physical quantities for fundamental frequency and time metrology – random instabilities. IEEE std. 1139–1999, 1999
5. Clock Jitter and Phase Noise Conversion, Application Note 3359, by Maxim. Available online at: www.maxim-ic.com
6. Overview on Phase Noise and Jitter, Agilent Technologies. Available online at: www.agilent. com
7. Hajimiri A, Limotyrakis S, Lee TH (1999) Jitter and phase noise in ring oscillators. IEEE J Solid-State Circ 34:790–804
8. Agilent Signal Generator Spectral Purity, Application Note 388. Available online at: www. agilent.com
9. Bottom VE (1981) A history of the quartz crystal industry in the USA. In: IEEE annual frequency control symposium, Philadelphia, PA, pp 3–12
10. Lam CS (2008) A review of the recent development of MEMS and crystal oscillators and their impacts on the frequency control products industry. In: IEEE ultrasonic symposium, Beijing, China, pp 694–704
11. Wan-Thai Hsu et al (2007) The new heart beat of electronics – Silicon MEMS oscillators. In: IEEE electronic components and technology conference, ECTC, Reno, Nevada, pp 1895–1899
12. SiTime's product selector sheet. Available online at: http://www.sitime.com/support/product-selector

13. McCorquodale MS et al (2010) A silicon die as a frequency source. In: IEEE international frequency control symposium, Newport Beach, California, pp 103–108
14. De Smedt V et al (2009) A 66 μW 86 ppm/ °C fully-integrated 6 MHz wienbridge oscillator with a 172 dB phase noise FOM. IEEE J Solid-State Circ 44(7):1990–2001
15. Gray PR, Hamilton DJ (1971) Analysis of electrothermal integrated circuits. IEEE J Solid-State Circ sc-6(1):8–14
16. Freidman MF, Hamilton DJ (1970) An integrated high-Q bandpass filter. In: IEEE international solid-state circuits conference, San Francisco, CA, February 1970, pp 162–163
17. Bosch G (1972) A thermal oscillator using the thermo-electric (seebeck) effect in silicon. Elsevier's Solid-State Electron 15:849–852
18. Wikipedia page on the thermoelectric effect. Available online at: http://en.wikipedia.org/wiki/Thermoelectric_effect
19. Szekely V et al (1995) A new monolithic temperature sensor: the thermal feedback oscillator. In: Proceedings of the transducers, STOCKHOLM, SWEDEN, June 1995, pp 124–127
20. Wikipedia page on phonons. Available online at: http://en.wikipedia.org/wiki/Phonon
21. Ebrahimi J (1970) Thermal diffusivity measurement of small silicon chips. J Phys D Appl Phys 3:236–239
22. Touloukian YS et al (1998) Thermophysical properties of matter, vol 10. Plenum, New York
23. Makinwa KAA, Snoeij MF (2006) A CMOS temperature-to-frequency converter with an inaccuracy of less than ±0.5 °C (3σ) from −40 °C to 105 °C. IEEE J Solid-State Circ 41(12):2992–2997
24. van Vroonhoven CPL et al (2010) A thermal-diffusivity-based temperature sensor with an untrimmed inaccuracy of ±0.2 °C (3σ) from −55 °C to 125 °C. In: IEEE ISSCC Dig. Tech. Papers, San Francisco, CA, pp 314–315
25. Wikipedia page on Moore's law. Available online at: http://en.wikipedia.org/wiki/Moore's_law
26. van Vroonhoven CPL et al (2008) A CMOS temperature to digital converter with an inaccuracy of ±0.5 °C (3σ) from −55 to 125 °C. In: IEEE ISSCC Dig. Tech. Papers, San Francisco, CA, pp 576–577
27. Pertijs MAP, Huijsing JH (2006) Precision temperature sensors in CMOS technology. Springer, Dordrecht

Chapter 2
Silicon-Based Frequency References

This chapter provides an overview of silicon-based frequency references. Reduction of size and cost as well as increased reliability have been the main motivations for the realization of on-chip frequency references. However, the main limitation of such references is the effect of variations in process, voltage, and temperature (PVT) on their output frequency. This chapter reviews various state-of-the-art implementations of silicon-based frequency references. These have been mainly introduced in the open literature or available as products on the market. The chapter's main goal is to provide an overview of the pros and cons of the selected approaches in order to build a comparison chart. Such overview should help the reader to make a comparison between the approach described in this book, electro-thermal frequency references, and the other available solutions.

2.1 Introduction

The stability of a frequency reference is a measure of the amount of variation in its output frequency as a function of environmental parameters. These include temperature, supply voltage, process tolerances, noise, etc. It should be noted that the terms stability and accuracy will be used interchangeably throughout this book. This is because they both refer to the same concept as far as the level of variations in the nominal oscillation frequency of an oscillator is concerned. If this nominal value is equal to f_0, then its level of stability (accuracy) is measured either in parts per million (ppm) or in percent [1–4]. If the absolute value of the deviation in the output frequency is Δf, then the error can be calculated as:

$$f_{error}(\%) = \frac{\Delta f}{f_0} \cdot 10^2 \quad \text{or} \quad f_{error}(ppm) = \frac{\Delta f}{f_0} \cdot 10^6. \tag{2.1}$$

S.M. Kashmiri and K.A.A. Makinwa, *Electrothermal Frequency*
References in Standard CMOS, Analog Circuits and Signal Processing,
DOI 10.1007/978-1-4614-6473-0_2, © Springer Science+Business Media New York 2013

Various electronic systems require different levels of accuracy for their frequency references. For instance, in some microcontroller applications references stable from 0.01% (100 ppm) to 1% (10,000 ppm) [5] might be required, while a wire-line data link such as USB 2.0 needs 500 ppm of clock accuracy [6]. Wireless communication channels require tighter accuracies. For instance a cell-phone handset application might need frequencies stable to 2.5 ppm [7], while a GPS receiver or a mobile base-station system might require sub-ppm accuracies [3, 7].

For decades, crystal oscillators have been the only means of producing stable frequencies. Considering their low temperature dependency, relatively low cost and small form factor, as well as their wide commercial availability, they have a dominant share of the frequency control market (more than 90%, equivalent to more than 4.5 billion dollars) [3]. Quartz crystal oscillators are available with various levels of accuracy. The non-compensated (XO) and voltage compensated (VCXO) oscillators achieve stabilities in the range of 20–100 ppm. When they are temperature compensated (TCXO), their accuracy is in the 0.1–5 ppm range. Oven controlled (OCXO) oscillators achieve very high stabilities: in the order of 1 ppb (part per billion) [3].

Apart from their high levels of accuracy, quartz crystal oscillators also have some drawbacks. The first of them is the space they occupy on printed circuit boards, especially when a number of frequency sources are required within one system. Another important disadvantage is their sensitivity to mechanical shock and vibration. This mainly affects the quartz crystal, which is in fact an electro-mechanical part [3]. Compared to electronic circuits, whose functionality is due to the movement of electrons, the crystal vibrates at the frequency of oscillation. This means that any physical motion of the crystal will change its frequency [3].

The abovementioned limitations have driven the search for integrated frequency references that can achieve the same level of stability as quartz crystal oscillators. Such references will be manufactured in silicon, which is why they are also referred to as silicon-based frequency references [8].

As early as 1967, the first steps towards frequency generation by means of MEMS (micro-machined silicon) structures were taken [9]. Around 1968, the concept of a self-referenced silicon frequency reference was illustrated with a temperature-compensated Wien-bridge RC oscillator. Later, various types of electrical oscillators such as RC, relaxation, ring, and LC oscillators have been proposed. Among these methods, MEMS-based and LC-based oscillators have been commercialized and currently achieve performance levels that can compete with crystal oscillators. In this chapter, an overview of these methods of silicon-based frequency generation will be described. State-of-the-art references will be studied in regard to their system-level architecture, their achieved level of accuracy, as well as an overview of their potential applications.

Since this book is about CMOS compatible frequency references, crystal oscillators will not be further discussed. Furthermore, MEMS-based oscillators, which are not truly standard CMOS compatible, will be briefly introduced in the next section. The chapter progresses with a more detailed overview of

CMOS-based LC, RC, relaxation, and ring oscillators. Furthermore, a new class of ultra-low-power frequency references based on the electron mobility of MOS transistors will be introduced. Finally, a comparison between these methods will be provided.

2.2 Silicon MEMS Based Oscillators

Quartz crystal resonators are excited at their resonance frequency by an electrical oscillator circuit. Their operation depends on the piezoelectric properties of a material that cannot be integrated in IC technology: quartz. Over the years, a lot of research has been done on the development of silicon MEMS (Micro Electro Mechanical Systems) based resonators with the aim of replacing quartz crystals. MEMS technology involves many of the processes used by the integrated circuit technology such as lithography, deposition, etching, etc. [10]. This technology has been applied in sensors such as accelerometers, gyroscopes, microphones, etc.

MEMS resonators are micro-machined structures that can vibrate at their resonance frequency if an external excitation is applied to them. The resonance property of such structures was first researched in 1967, when a resonant gate transistor was presented as a micro-machined integrated frequency reference [11]. This excitation can be of the electrostatic, piezoelectric or electromagnetic type [12, 13]. The *Quality* factor of a resonator determines the stability of the frequency reference that is built around it. It is the ratio of its peak resonance frequency to the width of the peak. A MEMS resonator's shape and geometry determines this factor, which is typically between 50,000 and 300,000, a range that is comparable to quartz crystal oscillators [12, 13].

MEMS resonators have faced many challenges in delivering a cost-effective and reliable solution that could compete commercially with quartz crystals. The major challenges included packaging, vibration and shock sensitivity, temperature drift and long term stability [14]. In recent years various commercial products have been introduced by two start-up companies: Discera and SiTime. Discera was established in 2001 based on research on MEMS resonators funded by DARPA, while SiTime started in 2004 based on IP licensed through Bosch [3]. Today, MEMS-based frequency references produced by these companies are more compact than their quartz competitors and are more cost effective due to the mass production allowed by the use of IC technology. However, their level of jitter (phase noise) is not (yet) low enough for cell-phone applications.

Because of the special processing required by MEMS technology, a MEMS resonator has to be manufactured on a separate die from the die that holds the electronic circuitry exciting and controlling it [12–14]. Furthermore, the mass of a MEMS resonator is small, being on the order of 10^{-14}–10^{-11} kg, which means that its resonance frequency and quality factor will be affected by any gas molecules surrounding it [15]. This means that silicon MEMS resonators should preferably be

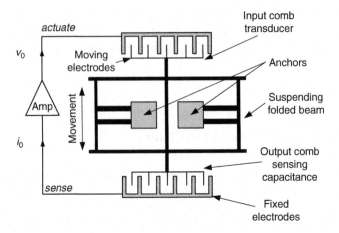

Fig. 2.1 Simplified block-diagram of a silicon MEMS based oscillator

operated in vacuum, which is the reason why they have been fabricated within silicon cavities [12–14].

Another challenge in making MEMS-based oscillators is the temperature dependence of MEMS resonators. This is due to the temperature coefficient of the Young's modulus of silicon [3, 12, 13]. This is in the order of 20–40 ppm/°C, which is larger than that of quartz and necessitates a means for the temperature compensation of such oscillators. There have been various structural techniques proposed to reduce or correct for the MEMS resonator's temperature coefficient. These include the combination of materials with positive and negative thermal stiffness coefficients or the application of an electric field to control the resonator's stiffness [16]. The approach that has been ultimately used in commercial products is to correct the temperature dependence of the oscillator through a fractional frequency synthesizer and a temperature sensor [17]. This technique will be described later.

An encapsulated silicon MEMS resonator needs to be attached to an anchor on a substrate [14]. Figure 2.1, shows a conceptual and simplified drawing of a MEMS resonator [18]. Folded suspending beams are anchored to the silicon substrate at two anchor points. The suspending beams are connected to the sides of comb transducer structures. The resonator structure is biased with a DC bias source. The output transducer experiences a change in capacitance due to the movement of the suspending beam with reference to the fixed electrodes. This causes an electrical signal, i_o, which is fed to an electronic circuit that produces an excitation signal v_o, which is then applied to the input transducer. This signal will electrostatically actuate the resonator. The structure vibrates at its resonance frequency (typically in the hundreds of kHz to MHz range), which is the same frequency at which it is excited electrically. The required electrical signal is in fact the output signal of the oscillator.

The MEMS frequency references produced by SiTime consist of a resonator element, which is wire bonded to a CMOS die that includes a sustaining circuitry,

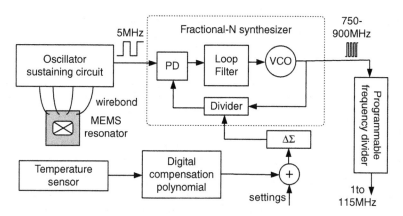

Fig. 2.2 Simplified block-diagram of a silicon MEMS oscillator, including a MEMS resonator, a fractional-N synthesizer and a temperature compensation scheme

a high-resolution fractional-N frequency synthesizer [19], a temperature sensor and digital circuitry [12–14, 17]. A simplified block diagram of this system [17] is shown in Fig. 2.2. The MEMS resonator vibrates at 5 MHz, which is the same frequency as that of the sustaining circuitry. This 5 MHz signal is provided to the fractional-N synthesizer, which outputs a higher frequency: in the range of 750–900 MHz [17]. This frequency can be adjusted with sub-ppm resolution over a 10% tuning range. A programmable output frequency can then be produced by dividing the output of the synthesizer. The advantage of this approach is that the same MEMS resonator can be used to provide different output frequencies. This means that the output frequency can be easily programmed into the device depending on the application.

The temperature dependence of the MEMS resonator is compensated by measuring the temperature of the CMOS die with an embedded temperature sensor. The temperature information is digitally processed through a compensation polynomial whose coefficients are stored in a non-volatile memory. The frequency reference achieves a part to part frequency stability of about 10 ppm from −40°C to 85°C [17]. In this approach, the jitter performance of the output frequency is determined by the frequency synthesizer (that is in principle a PLL). For better jitter performance, low noise and high quality factor oscillators such as LC based resonance circuits have been combined with optimized PLLs as well as power supply regulation techniques [13, 17].

One of the concerns regarding MEMS oscillators has been about their reliability in comparison to the mature quartz crystal rival. Since a MEMS resonator is a mechanical device that vibrates at millions of cycles per second, aging is one of these reliability concerns. Reliability tests published by Discera, show sub-ppm shifts in the first year of operation of such devices [20]. Furthermore, due to their very small dimensions (micro-meter range) and very small weight, MEMS resonators have better shock resistance than quartz crystals [20]. Further reliability tests such as vibration resistance, sensitivity to packaging vacuum, thermal cycling and high temperature storage life have been reported in [20], showing that MEMS frequency references can compete with crystal oscillators.

Most commercial MEMS frequency references are manufactured by SiTime [21] and Discera [22]. SiTime's high performance oscillators include the SiT8208, SiT8102 and SiT9102 in standard six-pin packages (5.0 × 3.2 mm^2), which are smaller than those currently used for quartz crystals [13]. SiTime also introduced very thin SiT8003 oscillators with 0.25 mm thick packages, mainly intended for SIM card, camera, and cell phone applications. SiTime's range of products cover output frequency stabilities from sub-ppm to 50 ppm over the commercial and industrial temperature range (−40°C to 85°C). The high performance SiT8208 and SiT8209 products have sub-ps output jitter [23].

Discera's MEMS frequency references use the same technique of combining a MEMS resonator with a PLL [22]. Their range of stability is about 50 ppm, at supply voltages of 1.8–3.3 V, output frequencies of 1–150 MHz and supply currents in the order of 3 mA. They are available in standard packages that can be placed in crystal oscillator footprints. Their intended applications are in: mobile applications, consumer electronics, portable electronics, CCD clocks for cameras, etc.

So far, the commercially introduced MEMS frequency references show that sub-ppm frequency stabilities and programmable output frequencies are feasible. Furthermore, their small footprints make it possible to replace standard crystals with MEMS-based devices. However, they still have a few drawbacks. Their jitter performance is determined by their fractional-N synthesizer and by the temperature compensation scheme. Also, the special processing required for the MEMS resonator makes single die integration of these devices difficult. This means that the integration of such frequency references as an IP block in a system-on-chip will usually result in a two-chip solution.

2.3 LC Oscillators

Another class of commercially available frequency references are the LC oscillators [24]. Such oscillators operate at the resonance frequency of an LC tank [25] and have been widely used in VCO's that produce RF range of frequencies [26]. These VCO's have been normally embedded into phase-locked loops (PLLs), with the aim of frequency synthesis from an external reference source. In order to function as a self-referenced frequency source, an LC oscillator needs to be free-running. In this case, special attention needs to be paid to its output frequency stability as a function of process, temperature and voltage variations. An LC oscillator is based on passive elements such as inductors and capacitors as well as active elements, i.e. transistors. Therefore, such an oscillator can be made in a standard CMOS process.

The first steps towards commercializing self-referenced LC oscillators were taken at Mobius Microsystems, a fab-less company founded in 2004 with the aim of developing all-silicon frequency sources that replace quartz crystal oscillators. The goal of Mobius Microsystems was to produce a monolithic free running RF LC

Fig. 2.3 Simplified block
diagram of an LC oscillator
including the LC elements as
well as their equivalent losses

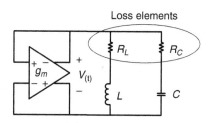

oscillator that did not require the frequency synthesizers used in MEMS frequency references. This was to avoid the effect of multiplication on the output frequency jitter. These efforts resulted in oscillators with output frequency ranges from 12 to 25 MHz and with initial target applications such as wire-line data communication, e.g. USB [27–32]. These solutions achieved output frequency stabilities in the order of 100 ppm with period jitters in the order of 3–6 ps (rms). In 2010, Mobius Microsystems was acquired by IDT, who has subsequently introduced LC oscillator based frequency references to the market [24].

A simplified block diagram of an LC oscillator is shown in Fig. 2.3. It includes an LC tank with inductor and capacitor values of L and C, respectively, each with their equivalent finite losses, R_L and R_C [31]. It also has a sustaining transconductor amplifier g_m (cross-coupled pairs) that compensates for the loss in the tank. The oscillation frequency is then [31]:

$$\omega = \frac{1}{LC}\sqrt{\frac{L - C \cdot R_L^2}{L - C \cdot R_C^2}}. \tag{2.2}$$

An LC oscillator based on the resonant tank shown in Fig. 2.3, not only suffers from frequency deviation due to the losses, but also due to variations in the absolute values of the passive elements due to process and temperature. The absolute values of integrated inductances have negligible temperature coefficient [31, 33], however, the temperature dependence of their equivalent loss resistance, R_L, is determined by the material from which the inductor is made. Since R_L is usually larger than R_C, the former's temperature dependence will be dominant. Furthermore, the capacitance will be affected by the fringing capacitors due to interconnect and parasitic capacitances of the transconductor g_m. The latter capacitance then has considerable temperature and bias dependence [31]. In principle, the temperature dependence of the output frequency of an LC oscillator shows a concave negative temperature coefficient, whose sensitivity increases at high temperatures [31].

The output frequency of an LC oscillator can also be affected if conducting materials are in its vicinity, since the field lines of the inductor will be affected by changes in the permeability or due to eddy currents [31, 32]. To overcome this problem, the solution proposed by IDT [32] is to build a Faraday shield around the die in order to maintain the fringing field lines and avoid disturbances. This is done by depositing a thick dielectric layer on the die of the LC oscillator chip, and

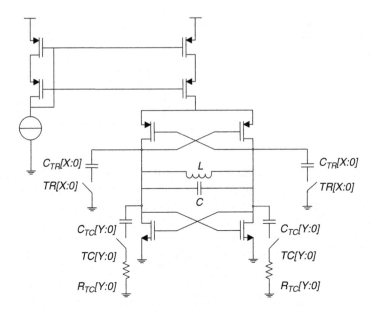

Fig. 2.4 Simplified circuit diagram of an LC oscillator and the trimming and temperature compensation networks

electroplating several microns of Cupper on top of that. The back side of the device is also shielded by means of an Aluminum layer [32].

A simplified circuit block diagram of the LC based oscillator used in the core of the frequency reference initially proposed by Mobius Microsystems and later turned into a product by IDT is shown in Fig. 2.4 [32]. In the initial publication [28] the resonance frequency was 1 GHz, which was later increased to 3 GHz [32] to increase the quality factor of the inductor. The LC oscillator consists of a cross-coupled negative transconductance amplifier with PMOS biasing transistors for low $1/f$ noise operation. An array of thin film programmable capacitors $C_{TR}[X:0]$ connected through the corresponding switches TR[X:0] are used to trim the output frequency. A set of thin film capacitors $C_{TC}[Y:0]$ and series resistors $R_{TC}[Y:0]$ can be connected through switches TC[Y:0], which are used to introduce a loss to the capacitive network. The type of RC network is chosen such that its temperature dependence works against that of the inductor's loss resistance to minimize the nonlinearity in the temperature coefficient of the oscillator [32]. The frequency reference includes a low drop-out regulator (LDO) to reduce the effect of power supply fluctuations, as well as a programmable divider allowing for programmable output frequency.

The LC oscillator introduced in [25] (0.35 μm CMOS) had an output frequency of 12 MHz and a supply current of 9.5 mA. It achieved a stability of about 400 ppm from −10°C to 85°C and a period jitter of <10 ps (rms). The work in [28] (0.25 μm CMOS) achieved a frequency stability of 90 ppm (shown for one device) from 0°C to 70°C. This work used an active temperature compensation scheme and dissipated about 15 mA. Its output jitter was <7 ps (rms). A major modification to these devices, towards reduction of their power consumption, was the change in their temperature compensation schemes. This initially included an active temperature compensation block including a PTAT generator and varactors in the LC tank, which were attached to a temperature dependent control voltage [29–31]. Later, this was changed to the passive temperature compensation scheme described in Fig. 2.4 [32]. As a result the supply current was reduced from 15 mA in [28] to less than 2 mA in [24, 32]. The frequency stability of these oscillators was about 300 ppm from 0°C to 70°C with a period jitter of 3.5 ps (rms). The recent product published by IDT [34] combines the previous active and passive temperature compensation schemes, adds an improved Faraday shield to solve the problem of interfering fields with the oscillator's resonance, and a two-point temperature trim to achieve a stability of <50 ppm from −20°C to 70°C and a sub-ps jitter level. Apart from the products introduced by IDT, the Si500 LC oscillators have been introduced by Silicon Labs [35]. These are capable of producing frequencies programmable from 0.9 to 200 MHz and are realized in a 0.13 μm CMOS. They draw 8 mA from a 1.8 V supply and are operational from 0°C to 70°C with stabilities in the order of 150 ppm.

Considering the number of temperature and process dependent variables in LC oscillators and the probable lack of correlation between these parameters, a single-point temperature trim is not sufficient for these devices. However, multiple point trims add to the production costs. Furthermore, LC oscillators have rather narrow temperature ranges, in the order of −20°C to 70°C, limiting their application operating over wide temperature ranges.

2.4 RC Harmonic Oscillators

The next class of oscillator reviewed here is the RC harmonic oscillator. This type of oscillator uses resistors and capacitors to form an RC network, which functions as a frequency selection circuit, and which can be combined with an amplifier to realize a linear oscillator with a sinusoidal output signal [36]. The output frequency of RC oscillators will be affected by variations in the absolute value of on-chip resistors and capacitors as well as their temperature dependence [37–39]. These variations can be in the order of tens of percent. By means of trimming and temperature compensation the stability of RC oscillators reaches about 1% [40]. Despite their lower accuracy, compared to the LC oscillators, RC oscillators are suitable for low frequency (hundreds of kHz to a few MHz) as well as low power (tens of micro-Watts) applications [40–43].

A well known type of harmonic RC oscillator is the Wien-bridge oscillator. This is based on an electrical network proposed by Max Wien in 1891 [40, 44]. As shown in Fig. 2.5, it includes two resistors and capacitors. The complete oscillator can be seen as a positive feedback amplifier with a band-pass network in its feedback path. In 1939, William Hewlett, a co-founder of Hewlett-Packard Company (HP),

Fig. 2.5 Simplified circuit
diagram of a Wien-bridge
harmonic RC oscillator

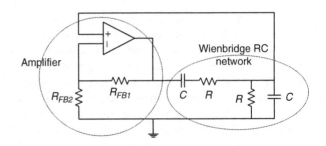

designed a Wien-bridge oscillator while an MSc student at Stanford University. This later led to HP200A, one of the first products of HP [45, 46], which was a low distortion oscillator for audio applications.

A remarkable point regarding the HP200A was its use of an incandescent bulb as a positive temperature coefficient thermistor in the oscillator's feedback path (R_{FB2}). This was for amplitude regulation. Without this, the output signal of the oscillator increases until it clips at the supply rails, thus creating harmonic distortion. The use of the bulb in the feedback path means that amplitude growth causes current increase, which heats the bulb, increasing its resistance and causing the current to decrease.

The Wien-bridge harmonic RC oscillator shown in Fig. 2.5 oscillates when the amplifier has a gain of 3 [40]. This gain is set by means of the resistive feedback network around the amplifier. At a gain of 3, the circuit will oscillate at:

$$f_{osc} = \frac{1}{2\pi RC}. \tag{2.3}$$

with R and C being the values of the elements in the passive feedback network. The network has a quality factor $Q = 1/3$ [40]. To first order, the process and temperature stability of this frequency are determined only by that of the passive R and C elements. In the implementation reported in [40], metal-insulator-metal (MiM) types of capacitors have been used, whose temperature dependence is reported to be negligible. In addition, positive temperature coefficient N-poly resistors were combined with negative temperature coefficient P-poly resistors, leading to a residual temperature coefficient of 36 ppm/°C. The remaining source of spread is the variation in the absolute values of these elements, which can be up to 10%, requiring a process trim to be applied to the oscillator.

Other non-idealities contributing to the inaccuracy of the output frequency are related to the finite gain, output impedance and the phase shift introduced by the amplifier. In order to mitigate their effects, the fully differential and modified Wien-bridge oscillator circuit shown in Fig. 2.6 has been proposed [40, 41]. Transistor T_1 forms the amplifier, which is degenerated by R_{deg} and cascoded by gain-boosted cascode transistors T_2 and T_3 (gain-boosters not shown), and biased with extra current-bleeding sources I_b. Resistor degeneration has been used to guarantee that the amplifier's gain of 3 is defined by the ratio R/R_{deg}. To enhance the degeneration by maximizing the transconductance of T_1 and without sacrificing

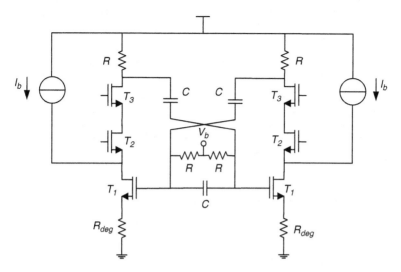

Fig. 2.6 More detailed circuit diagram of a Wien-bridge RC oscillator

output impedance, current bleeding has been applied. The gain-boosted cascode transistors further increase the output impedance such that it does not interfere with the Wien-bridge network's transfer function. To minimize excess phase shift, the cascode transistors are minimum size devices.

The harmonic Wien-bridge RC oscillator of [40] was implemented in a 65 nm CMOS process and dissipates 55 μA from a 1.2 V supply. With $R = 50\text{k}\Omega$, and $C = 530\text{fF}$, it oscillates at about 6 MHz. Its measured absolute accuracy (for six devices characterized from 0°C to 120°C) is reported to be 0.9%, with a temperature coefficient of about 86 ppm/°C.

Another Wien-bridge oscillator based on [40] was proposed in [43], where the RC oscillator circuit is combined with a low drop out voltage regulator in order to achieve a supply dependency of 104 ppm/V. The proposed circuit oscillates at 24 MHz and was implemented in 0.13 μm CMOS. It operates from a supply voltage range of 0.4–1.4 V and dissipates 37 μA. No measurement results were reported on its performance over temperature.

2.5   RC Relaxation Oscillators

Another type of RC-based oscillator is the relaxation oscillator. This produces a digital (square-wave) output signal [47]. Its output frequency can range from hundreds of kHz to a few tens of MHz. In this case, the period of oscillation is proportional to a time constant, determined by a reference voltage, a bias current I_{ref} and a capacitor C (see Fig. 2.7). Since the bias current typically involves a

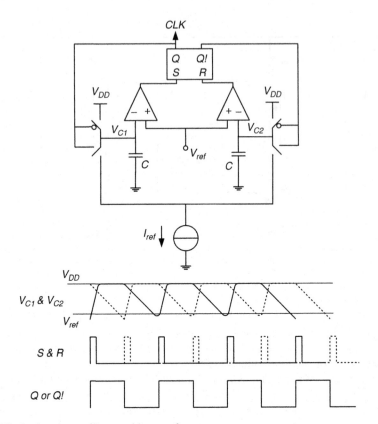

Fig. 2.7 A relaxation oscillator and its waveforms

resistance (e.g. by forcing a band-gap reference voltage across a resistor), they are also recognized as RC oscillators.

CMOS compatibility and very low power operation are some of the major advantages of relaxation oscillators. These characteristics make them suitable for battery powered applications such as the wake-up timers in implantable biomedical systems [48–50]. However, one of their main drawbacks is the dependence of their output frequency on process and temperature variations [48, 49]. This is mainly limited to about 20% by the tolerance and temperature dependence of conventional on-chip resistors and capacitors. By means of temperature compensation and trimming, stabilities of 2% have been achieved [49, 50], sufficient for the previously mentioned applications.

One possible realizations of a relaxation oscillator is shown in Fig. 2.7. This circuit includes a current and a voltage reference, two capacitors and comparators as well as a set-reset (SR) latch. Its operation involves charging capacitors C_1 and C_2 to V_{DD} and then discharging them to V_{ref} by means of I_{ref} (see the waveforms in Fig. 2.7). The Schmitt trigger comparators compare the capacitor voltages with V_{ref}

and change the state of the SR latch, which, in turn, changes the charge and discharge order of the capacitors [50]. The oscillation frequency is determined by:

$$1/f_{osc} = T_{osc} = \frac{C}{2 \cdot I_{ref}} \left[2 \cdot \left(V_{DD} - V_{ref} \right) \right]. \tag{2.4}$$

The flicker ($1/f$) noise of the current reference I_{ref}, as well as the input referred noise of the comparators both contribute to the output jitter. Furthermore, the input referred offset and the process and temperature dependent delay of the comparators will influence the oscillator's accuracy. Various solutions to these issues have been addressed in [50–53].

The designs described in [48, 49] are based on the circuit shown in Fig. 2.7. The reference current and voltage are based on a band-gap reference generator and an 8-bit digital trim can be applied to the oscillator. Without trimming, the reported accuracy of the 12 MHz output frequency is about ±25%, which after trimming (single point) remains stable to ±5% over supply variations and a temperature range of −40°C to 125°C. A 0.5 μm CMOS realization reported in [48] dissipates about 3 μW and its output jitter is in the order of 0.1%. The realization in [49] improves on [48] by reducing the inaccuracy to ±2.5% under the same conditions, which is achieved by correcting for the residual temperature coefficient of the oscillator after the trimming.

The relaxation oscillator presented in [50] is intended for low power biomedical applications. The oscillator is implemented in a 0.13 μm CMOS process, and produces an output frequency of 3.2 MHz. It dissipates 38.4 μW from a 1.5 V supply. The oscillator is also based on the topology shown in Fig. 2.7, with the application of an auto-zeroing mechanism [51] to the oscillator's comparators. This reduces their offset and flicker noise, which improves the oscillator's accuracy and reduces its output jitter, respectively. An 8-bit digital trim of the reference current of the oscillator has a resolution of 0.3%. The reported variation in the output frequency over a temperature range of 20–60°C is ±0.25%. The oscillator's cycle-to-cycle jitter is 524 ps (rms) without the application of auto-zeroing, which drops to 455 ps (rms) when the comparators are auto-zeroed.

A voltage-controlled-oscillator (VCO) based on a relaxation oscillator is presented in [52], with the aim of achieving improved control linearity and jitter performance. An alternative Schmitt trigger circuit is introduced as the oscillator's comparator. The oscillator dissipates 360 μA and has a jitter of 65 ppm (rms) at an oscillation frequency of 1.5 MHz. Its control linearity is affected by the comparator delay, which needs to be minimized compared to the period of oscillation. This could be solved by increasing the comparator bandwidth at the cost of increased noise and jitter [52]. However, in [52] the direction of the capacitor's charging current is changed gradually rather than instantaneously. The accuracy of this oscillator is not reported in [52].

The 12.5 MHz relaxation oscillator described in [53] is implemented in a 65 nm CMOS process and has a current consumption of 70 μA with a 1.2 V supply. This work focuses on the reduction of comparator noise to reduce the total oscillator's

Fig. 2.8 A relaxation oscillator with voltage-averaging feedback

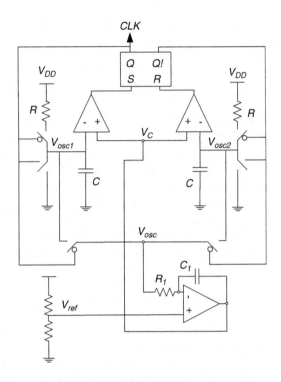

phase noise and jitter. Subtraction of charge by a switched capacitor circuitry filters the comparator noise, which then allows its power consumption to be significantly reduced. The measured phase noise of this oscillator is -82dBm at an offset frequency of 100 kHz. There is no characterization of the temperature stability of this oscillator reported in the corresponding publication.

As described earlier, one of the variables contributing to the output frequency variation of a relaxation oscillator is the delay of its comparator. This is the time it takes the comparator to change the state of the latch after the capacitor voltage reaches V_{ref} (see Fig. 2.7). This delay forms part of the oscillation period, and increases its spread as a function of PVT. This delay can be minimized in comparison to the oscillation period, but at the cost of excess power consumption. The work in [54] proposes a voltage-averaging feedback loop that makes the oscillation period insensitive to the comparator delay.

The voltage-averaging feedback topology [54] is shown in Fig. 2.8. The core of the oscillator can be seen as a relaxation oscillator with two comparators and a control voltage V_C, which is tuned by the voltage averaging feedback circuitry. The two internal voltages V_{osc1} and V_{osc2} (see Fig. 2.9) are alternately applied to the active filter in the voltage averaging feedback circuitry. The active filter made of resistor R_1, capacitor C_1 and an opamp ensures that at all conditions the DC value of V_{osc} is equal to V_{ref}. This means that these two voltages are virtually shorted in a low-frequency bandwidth determined by $R_1 \cdot C_1$ (about 160 kHz with $R_1 = 1\text{M}\Omega$ and $C_1 = 1$ pF). Assuming an ideal opamp in the feedback circuit and

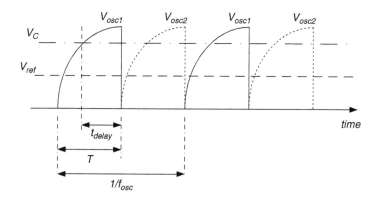

Fig. 2.9 Waveforms of the voltage-averaging relaxation oscillator

considering that the main oscillator capacitor C is being charged to V_{dd} through a resistor R $(R_1 >> R)$ the waveforms shown in Fig. 2.9 can be modeled as:

$$V_{osc1,2}(t) = V_{dd}\left(1 - e^{-1/RCt}\right). \tag{2.5}$$

Due to the voltage averaging feedback, the DC value of $V_{osc1,2}$ needs to be equal to V_{ref} in half of an oscillation period (T):

$$\frac{1}{T}\int_0^T V_{osc1,2}(t)dt = V_{ref}. \tag{2.6}$$

As shown by Fig. 2.8, the reference voltage V_{ref} is made by a resistive divider from V_{dd} with a division factor $\alpha = V_{ref}/V_{dd}$. Therefore, (2.6) results in [54]:

$$\frac{(1-\alpha)T}{RC} = 1 - e^{-T/RC}. \tag{2.7}$$

Which means that the oscillation period is ideally only defined by R, C and α. Therefore, variations in comparator delay will have no effect on the oscillation period, because the voltage averaging feedback loop controls V_C to keep the frequency constant. Furthermore, (2.7) shows that the dependence of oscillation period on supply voltage is cancelled. Finally, the low-pass nature of the voltage averaging network (active integrator in the feedback) means that the low frequency (flicker) noise referred to the input of the oscillator will be high-pass filtered, which results in a reduction of the output jitter [54]. The 0.18 µm standard CMOS oscillator presented in [54] dissipates 25 µA from a 1.8 V supply. The reported accuracy of the oscillator's 14 MHz output frequency is ±0.19% (result of a single device) from −40°C to 125°C, while its cycle-to-cycle jitter is 30 ps (rms).

2.6 Ring Oscillators

Ring oscillators are widely used as voltage-controlled oscillators in jitter sensitive applications such as phase-locked loops and clock recovery circuits. This is due to the high frequencies that they can achieve and their relatively simple integration [55]. Ring oscillators are widely realized in CMOS process as a ring of cascaded inverter stages [56]. The number of inverters needs to be odd and the output of the last stage has to be fed back to the input of the first stage (see Fig. 2.10). For an odd number of stages, the output of the last stage is the inverse of the input of the first stage. Propagation delay of the cascaded stages delays the output of the last stage compared to the input of the first stage. This results in the oscillation to propagate in the ring. A half period of oscillation will be equal to the number of inverter stages times the delay of each stage.

The inverter stages could also be made by means of analog delay stages such as the ones published in [57, 58] and shown in Fig. 2.11. This fully differential delay stage is made of a differential pair and a symmetrical load. The time delay introduced by this stage is approximated by [59]:

$$t_d \approx \frac{C_0(V_H - V_L)}{I_{ref}}. \tag{2.8}$$

Where C_0 is the total capacitance at the output of the stage, I_{ref} is the bias current of the circuit and $V_H - V_L$ is the output voltage swing, determined by V_{ref} and V_{CTRL}

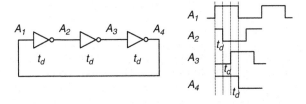

Fig. 2.10 Simplified ring oscillator schematic and its conceptual timing diagram

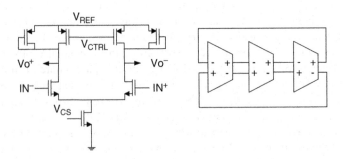

Fig. 2.11 Differential delay cell and a three stage oscillator

[57]. The ring oscillator proposed in [57] is made of a cascade of three delay stages (see Fig. 2.11). To vary the time delay t_d, and thus the output frequency, the control voltage V_{CTRL} can be modified.

The delays of the stages of a ring oscillator vary as a function of PVT. Supply voltage variations affect the voltage domain parameters in (2.8), while the temperature dependence of MOS transistors and passive components affects the reference currents and voltages. In order to reduce these effects, two main approaches can be found in literature. The first approach takes free running and open-loop ring oscillators and compensates the PVT effects within the circuit. The second approach embeds a voltage/current controlled ring oscillator in a feedback loop. As a result of feedback, the output frequency of the ring oscillator is locked to the time constant produced by a temperature compensated resistor and capacitor network. These two approaches will be discussed in the following sub-sections.

2.6.1   Open-Loop Compensation

The work presented in [57] describes how PVT compensation techniques can be applied to a free running, open-loop ring oscillator. An earlier oscillator, presented in [58], had a nominal frequency of 680 kHz and was implemented in a 0.6 μm CMOS process with inaccuracy of ±6.8% from 35°C to 115°C. Later, an improved implementation in a 0.25 μm CMOS process was presented in [57] with an output frequency of 7 MHz. The improved oscillator had an untrimmed inaccuracy of ±2.6% over process, supply and a temperature range of −40°C to 125°C. The reported accuracy is based on the data collected from 94 samples from two different batches. A block diagram of the complete system, including the process and temperature compensation as well as the supply regulation around the core oscillator, is shown in Fig. 2.12.

A band-gap voltage reference regulates the variable supply (2.4–3 V) to a supply and temperature stable 2.2 V reference voltage V_{ref}. This voltage is then used by other blocks in the system. The reference frequency is produced by the differential three stage ring oscillator at the core of the system, whose frequency is stabilized by a reference current I_{ref} (see Fig. 2.11). This current is derived from the control voltage V_{CTRL}, which is generated by the temperature and process compensation circuitry. The analog output of the ring oscillator is translated by a rail-to-rail swing comparator to digital voltage levels.

The control voltage V_{CTRL} is varied to correct for temperature and process variations. The critical temperature dependent parameters are the mobility, $\mu_{P,N}$, and the threshold voltage, $V_{THP,N}$, of the MOS transistors [60]. The mobility has an approximate $T^{-1.5} \sim T^{-2.2}$ temperature dependence, where T is the absolute temperature. The threshold voltage however has a negative temperature coefficient. Furthermore, the junction capacitances of a MOS device as well as the oxide capacitance have temperature dependences [57, 60]. Process variations affect gate oxide thickness and doping concentrations, leading to threshold voltage and

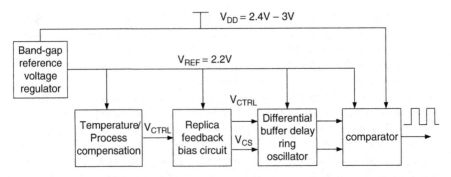

Fig. 2.12 System block diagram of the temperature and process compensated ring oscillator

mobility variations. In [57] a bipolar transistor's base–emitter voltage V_{BE}, which has a negative temperature coefficient [61], was used to correct for the overall negative temperature coefficient of the ring oscillator. Furthermore, a threshold voltage detection circuit detects the process corner and modifies the reference voltage of the oscillator accordingly.

2.6.2 Closed-Loop Compensation

A ring oscillator with temperature compensation feedback loop is described in [62]. The oscillator has an output frequency of 10 MHz, dissipates 80 μW, and has a temperature dependence of 67 ppm/°C. This supply regulated oscillator, has been implemented in a 0.18 μm CMOS process. Its frequency stability is ±0.4% from −20°C to 120°C (data reported for a single device). There is no data available on the jitter performance of this oscillator.

A block diagram of the closed loop oscillator of [62] is shown in Fig. 2.13. The core ring oscillator has four differential stages and its supply, V_{CTRL}, is regulated by a feedback loop. This locks the output frequency of the ring oscillator to a PVT insensitive voltage V_{ref}. To close the feedback loop, a frequency-to-voltage converter (shown in Fig. 2.13) is used, which also realizes a linear temperature compensation scheme. To further investigate the operation of the loop, the timing diagram shown in Fig. 2.14 needs to be considered together with the block diagram of Fig. 2.13.

A voltage regulator based on a band gap reference produces a 1 V voltage V_{REG} from the variable V_{DD} (1.2 ∼ 3 V). When the reset signal RST is high, the charge on capacitor C_0 is set such that $V_{cap} = V_{REG}$. After the reset phase, the frequency conversion phase begins with signal Q = 0, discharging C_0 by current I_{ref}. The signal Q is then produced through a frequency divider by dividing the oscillator output frequency by two (see Fig. 2.13). The discharge of C_0 continues until signal Q changes state again. This always happens at the end of the oscillation period. At the rising edge of Q, and for any given value of I_{ref} and C_0, the voltage across

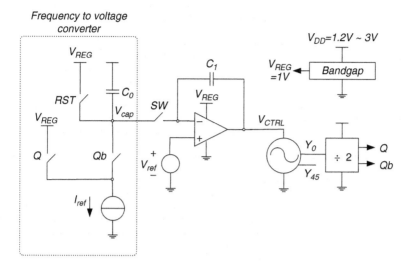

Fig. 2.13 Ring oscillator with a frequency-to-voltage converter and feedback loop

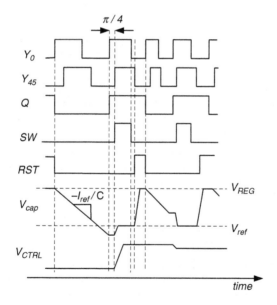

Fig. 2.14 Timing diagram of the ring oscillator in a feedback loop

C_0, V_{cap}, will be a function of the oscillation period. During the time SW $= 1$ (both SW and RST are generated by a digital control block that is not shown in the figure), V_{cap} is compared to a reference voltage V_{ref} through a switched capacitor loop filter. This works based on charge transfer from C_0 to C_1, which leads to a change in V_{CTRL}. Due to feedback around the integrator, the voltage V_{cap} will be forced to be

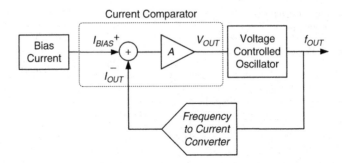

Fig. 2.15 Block diagram of the ring oscillator with frequency to current converter

equal to V_{ref}. Since V_{cap} is a representative of the oscillation period, for a constant V_{ref}, having $V_{cap} = V_{ref}$, is equivalent to having a constant output frequency. At steady-state, the output frequency, f_{CLK}, can be derived from [62]:

$$f_{CLK} = \frac{I_{ref}}{2C_0(V_{REG} - V_{ref})}.$$ (2.9)

To ensure a stable output frequency, the values of V_{REG}, V_{ref}, and I_{ref} need to be insensitive to PVT. The first two are made by means of a band-gap voltage reference and a sub-threshold voltage divider, respectively [62].

Another ring oscillator embedded in a control feedback loop is presented in [63]. This is an implementation in a 0.35 μm CMOS with a tunable output frequency from 2 to 100 MHz. The oscillator dissipates 180 μW at 30 MHz and has a process sensitivity of 2.7% and a temperature coefficient of 90 ppm/°C. A simplified block diagram of this oscillator is shown in Fig. 2.15. The circuit includes a bias current generator circuit, a current comparator, a ring oscillator based VCO and a frequency-to-current converter arranged in a frequency-locked loop. The current comparator block produces the output voltage V_{OUT} driving the VCO based on the difference between I_{BIAS} and I_{OUT}. Feedback forces these two currents to be equal.

The complete circuit diagram of the oscillator is shown in Fig. 2.16. It consists of a bias current generator that produces a bias current I_{BIAS} from the series combination of positive and negative temperature coefficient resistors. This is done by copying the bias voltage V_{BIAS} through a feedback amplifier to the resistors. Furthermore, the circuit includes a current comparator made of a simple common-source stage comparing I_{BIAS} and I_{OUT} (the output current of the frequency-to-current converter). The output of this stage is the control voltage V_{OUT}. The core ring oscillator is made of seven current-starved inverter stages. Its oscillation frequency, f_{OUT}, is determined by the applied current I_b to the inverters [63]:

$$f_{OUT} \propto \frac{I_b}{2mC_LV_{DD}}.$$ (2.10)

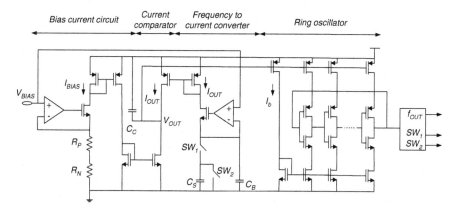

Fig. 2.16 Circuit diagram of ring oscillator with frequency-to-current converter in a feedback loop

Where m is the number of inverters and C_L is each inverter's load capacitance. The variations in control voltage V_{OUT} determine the current I_b and thus the oscillation frequency.

The frequency to current converter is made of a switched capacitor resistance made of capacitor C_S ad two switches $sw_{1,2}$ that are driven by the output of the oscillator. This means that the switched capacitor resistance is proportional to the oscillation frequency, and therefore, by copying the V_{BIAS} voltage onto this resistor, a frequency dependent output current I_{OUT} will be produced [63]:

$$I_{OUT} = f_{OUT} \cdot C_S \cdot V_{BIAS}. \tag{2.11}$$

A current mirror copies this current to the current comparator. At the steady state of the frequency-locked loop, the oscillation frequency will be given by:

$$f_{OUT} = \frac{1}{(R_P + R_N) \cdot C_S}. \tag{2.12}$$

Where R_P and R_N are the positive and negative temperature coefficient resistors used to produce the I_{BIAS} current from V_{BIAS}. To first order, the residual temperature coefficient of the composite resistor as well as that of the capacitor C_S determines the temperature coefficient of the oscillator.

For the closed-loop ring oscillators discussed so far, the oscillator's frequency is locked to a time constant defined by an RC circuit. These circuits provide a continuous correction of the frequency in response to supply voltage and temperature variations. Their achieved level of PVT variation is determined by that of the RC circuit.

2.7 Mobility-Based Frequency References

Recently, a class of low-power temperature-compensated frequency references based on the mobility of MOS transistors has been introduced [64–66]. Such references dissipate micro watts of power and achieve inaccuracies in the order of a few percent. The implementation described in [65] has been targeted for wireless sensor networks. It has an output frequency of 150 kHz, dissipates 42.6 μA from a 1.2 V supply, and is fabricated in a 65 nm standard CMOS process. For a temperature range of −55°C to 125°C, the reference achieves an output frequency stability of ±0.5% when trimmed at two temperature points. With a single trim its inaccuracy is ±2.7%. Another mobility based oscillator reported in [66] has been implemented in a 0.35 μm CMOS, has an output frequency of 3.3 kHz, and consumes 11nW from a 1 V supply.

The temperature dependence of mobility is in the order of $T^{-1.6}$ [65], which means that temperature compensation needs to be applied to a mobility based frequency reference. The core of the reference proposed in [65] is a current-controlled relaxation oscillator, which is controlled by a current that is proportional to the electron mobility. The temperature compensation of the oscillator is performed digitally (see Fig. 2.17). A band-gap temperature sensor [67] measures the temperature of the die. Through a non-linear digital mapping, the sensor's digital output is translated into a temperature-dependent division factor N_{div}, which is then applied to a divider. This divides the oscillator's output frequency by N_{div} in order to produce a stable output frequency f_{osc}.

A simplified circuit schematic of the mobility-based oscillator as well as its timing diagram is shown in Fig. 2.18. The voltage difference between the gates of M_1 and M_3 is kept equal to V_R by the combination of the current source I_0, R_0 and OA_1 [65]. Using the square-law MOS model, the drain current of M_1 is determined by:

$$I_1 = \frac{\mu_n C_{ox}}{2} \frac{W_1}{L_1} \frac{V_R^2}{\left(\sqrt{\frac{n}{m}} - 1\right)^2}. \tag{2.13}$$

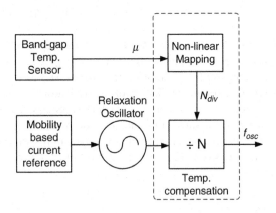

Fig. 2.17 Simplified block diagram of the temperature compensated mobility based frequency reference

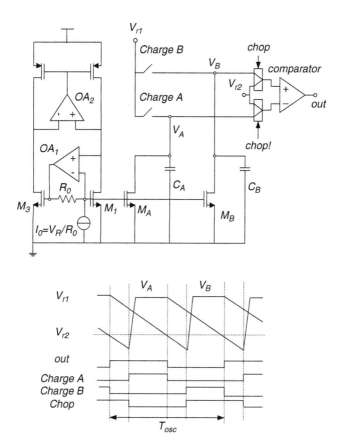

Fig. 2.18 Circuit and timing diagrams of the mobility based frequency reference

where m = $(W/L)_3/(W/L)_1$ and n = $(W/L)_4/(W/L)_2$, μ_n is the electron mobility, C_{ox} is the oxide capacitance of the MOS transistor per unit area. The current source I_0 is implemented by mirroring the current flowing in a resistor matched to R_0 whose voltage drop is equal to the reference voltage V_R (not shown in the schematic).

The drain current of M_1 is mirrored by M_A and M_B with a gain of four and used to alternatively discharge C_A and C_B after they have been pre-charged to V_{r1}. A comparator changes the state of charge and discharge of capacitors when the voltage on the discharging capacitor drops below V_{r2}. The oscillation frequency of this oscillator can be derived from [65]:

$$f_{osc} = \frac{\mu_n C_{ox}}{4C\left(\sqrt{\frac{n}{m}}-1\right)^2} \cdot \frac{W_1}{L_1} \cdot \frac{V_R^2}{V_{r1}-V_{r2}}. \tag{2.14}$$

where C = C_A = C_B $\propto$ C_{ox}.

A band-gap temperature sensor based on a bipolar core including NPN transistors and a first order delta-sigma ADC produces a digital representation of the die temperature [67]. The temperature sensor's accuracy should not limit the compensated oscillator's output frequency stability. For a frequency error of about 0.3%, the temperature sensor's accuracy needs to be better than 0.5°C [65]. A division factor N_{div} with 13-bits resolution has been chosen. A seventh order polynomial with fixed coefficients is used to translate the output of the temperature sensor to the division factor N_{div}.

Another relaxation oscillator locked to the mobility of MOS transistors has been implemented in a 0.35 μm CMOS process and is reported in [66]. This oscillator has an output frequency of 3.3 kHz and consumes 11nW from a 1 V supply. Its reported temperature drift from −20°C to 80°C is 500 ppm/°C, which accounts for a total temperature variation of about 5%. The key features in this topology are the generation of an electron mobility based reference current and the production of an oscillator reference voltage with the same temperature dependence as the electron mobility, to temperature compensate the output frequency. Further power reduction has been achieved by using a relaxation oscillator based on a single comparator and reducing the power consumption of the digital circuitry by means of a reduced supply voltage operation through a voltage regulator. The reported process spread of this oscillator, measured at a single temperature point, amounts to about ±15% [66].

2.8 Comparison

This chapter provided an overview of the various types of silicon-based frequency references. These included MEMS-resonator-based oscillators, LC oscillators, RC harmonic oscillators, RC relaxation oscillators, ring oscillators and finally electron-mobility-based oscillators. Each approach has its own specific advantages and disadvantages. To make a comparison, it is helpful to summarize the performance characteristics for each type of frequency reference.

One of the difficulties in providing a complete and fair comparison between published frequency references is that often insufficient data on their performance over process and temperature has been provided. There are references in which the performance of a single device has been reported as a measure of stability, which does not allow a fair comparison with references for which more samples have been characterized. From the previously described types of silicon-based frequency references, a performance summary of those with the most complete results is presented in Table 2.1.

MEMS-resonator-based oscillators and LC oscillators have been commercialized, and thus their reported performance characteristics are at production level. The characteristics of other topologies are obtained mainly from publications. It can be seen that MEMS-based oscillators achieve the best accuracy over the widest temperature range. Their form factor has also been shrunk and they

Table 2.1 Comparison of some state-of-the-art all-silicon frequency references

Reference number	[23]	[32]	[43]	[57]	[65]	[54]
Reference's principle of operation	MEMS	LC	RC harmonic	Ring	Mobility	Relaxation feedback
Frequency range	1–800 MHz	24 MHz	6 MHz	7.03 MHz	150 kHz	14 MHz
Temperature range (°C)	−40 to 125	0–70	0–120	−40 to 125	−55 to 125	−40 to 125
Supply voltage (V)	1.8–3.3	1.8	1.2	2.5	1.2	1.8
Power consumption (or supply current)	3.2–20 mA	<4 mW	66 µW	1.5 mW	51 µW	45 µW
Area (mm²)	Not available	0.8	0.03	1.6	0.2	0.04
Process	MEMS + CMOS	0.13 µm	65 nm	0.25 µm	65 nm	0.18 µm
Accuracy (ppm)	20–100	±300	±9,000	±18,400	±5,000 (two point trim) and ±27,000 (single trim)	±1,900
Temperature coefficient (ppm/°C)	Not available	±8.6	±86	±50.9	Not available	±23
Period jitter (rms)	<5 ps	<2 ps	Not available	Not available	52 ns	30 ps
Number of samples reported	Commercial	Commercial	6	94	12	1

are physically smaller than crystal oscillators. However, their major disadvantage is their need for special MEMS processing. This requires a two-die solution, in which the MEMS resonator is wire bonded to another CMOS chip. The power consumption of MEMS oscillators is comparable to LC oscillators and is larger than the other types.

Apart from the MEMS-based oscillators, all the other frequency references in Table 2.1 are standard CMOS compatible, which is a great advantage as far as manufacturing and packaging costs and complexity are considered. Among these, LC oscillators achieve the best accuracy over process and temperature, as well as the best jitter performance. However, their power consumption is higher than the rest and their temperature range is the narrowest. MEMS-based and LC-based oscillators are the only solutions that can achieve accuracies better than 0.1% at a reasonable jitter level.

For less accurate applications with stability requirements above 1% and with stringent power consumption requirements, RC, mobility-based, and relaxation oscillators can be used. These oscillators have very low chip area and can operate at the micro-Watt range. They are well suited for battery powered applications such as wireless sensor networks or biomedical implants.

2.9 Conclusions

Most electronic devices require a frequency reference. The stability of this reference, i.e. its deviation as a function of PVT, is crucial for many applications. For decades, crystal oscillators have been the dominant frequency control components. So far, they have achieved the best stability and noise/jitter performance; however, they have a few drawbacks. Their integration in the IC technology is nearly impossible. Furthermore, they are sensitive to shock and vibration, which can cause reliability problems.

In order to replace the crystal oscillators, there has been a tremendous R&D effort over the past few years. The goal is to produce integrated frequency references that can achieve the stability of crystal oscillators. This is justified by the large market for frequency generation and control components. Furthermore, an integrated, or in other words all-silicon, frequency reference brings added-value in terms of reliability and reduced area and cost.

One of the successfully commercialized all-silicon frequency references is the MEMS resonator based oscillator. This requires a two-die solution that combines a MEMS resonator with a CMOS chip. MEMS oscillators already achieve levels of stability below 1 ppm, which is comparable to that of crystal oscillators. Their stability is not altered by shock and vibration and their form factor allows for their placement in the same foot print as the crystal oscillators.

It will be still more advantageous if an all-silicon frequency reference could be made that is standard CMOS compatible. This would allow for cheap and simple production and packaging, as well as integration in systems-on-chip. So far, such references have been based on time constants produced by means of on-chip passive elements such as resistors, capacitors and inductors.

Among such CMOS compatible oscillators, the LC type has so far become commercial with stabilities in the order of a few tens of ppm's. This is achieved by means of trimming and temperature compensation. The RC type oscillators have stabilities in the order of 1%; however, their power consumption is much lower than that of LC oscillators, allowing for their use in battery powered applications. Furthermore, temperature compensated oscillators locked to the electron mobility of MOS transistors have been recently introduced achieving stabilities in the order of 1%.

The all-silicon oscillators reviewed so far in this chapter are based on signals which are produced, transferred and processed in the mechanical, electrical and magnetic energy domains. This is mainly done through the use of electromechanical structures or electrical passive components in combination with electronic circuitry. The main focus of this book is on investigating the possibility of accurate on-chip frequency generation through the generation and transfer of signals in another physical energy domain: the thermal domain. The ultimate goal is to make an accurate CMOS-compatible frequency reference. In the following chapters, the design of an electrothermal frequency reference based on the thermal properties of IC grade silicon will be investigated.

References

1. Bottom VE (1981) A history of the quartz crystal industry in the USA. In: IEEE annual frequency control symposium, Philadelphia, Pennsylvania, pp 3–12
2. Allan D et al (1997) The science of timekeeping, HP Application Note 1289
3. Lam CS (2008) A review of the recent development of MEMS and crystal oscillators and their impacts on the frequency control products industry. In: IEEE ultrasonic symposium, pp 694–704
4. McCorquaodale MS et al (2009) On modern and historical short-term frequency stability metrics for frequency sources. In: IEEE frequency control symposium, pp 328–333
5. Oscillators for microcontrollers, Intel Application Note AP-155, 1983
6. Universal Serial Bus (USB) Specifications Rev 3.0, 2008. Available online at: www.usb.org
7. Allan DW et al (1988) Ensemble time and frequency stability of GPS satellite clocks. In: IEEE annual frequency control symposium, pp 465–471
8. McCorquodale MS (2009) Silicon challenges quartz: precision self-referenced solid-state oscillators for frequency control and generation. In: IEEE Toronto section, University of Toronto, Canada, 2009. Available online at: www.toronto.ieee.ca/chapters/ssc/mccorquodaleUToronto09.pdf
9. Nathanson HC et al (1967) The resonant gate transistor. IEEE Trans Electron Dev 14 (3):117–133
10. Sadiku M (2002) MEMS. IEEE Potential 21(1):4–5
11. Nathanson HC et al (1965) A resonant-gate silicon surface transistor with high-Q bandpass properties. IEEE Trans Electron Dev 12(9):507
12. Tabatabaei S et al (2010) Silicon MEMS oscillators for high-speed digital systems. IEEE Micro 30(2):80–89
13. MEMS replacing quartz oscillators, SiTime Application Note AN10010, 2009

14. Lutz M (2007) MEMS oscillators for high volume commercial applications. In: IEEE transducers, pp 49–52
15. Wan-Thai Hsu et al (2007) The new heart beat of electronics - Silicon MEMS oscillators. In: IEEE electronic components and technology conference, ECTC, pp 1895–1899
16. Ruffieux D et al (2010) Silicon resonator based 3.2 uW real time clock with 10 ppm frequency accuracy. IEEE J Solid-State Circ 45(1):224–234
17. Perrott MH et al (2010) A low-area switched-resistor loop-filter technique for fractional-N synthesizers applied to a MEMS based programmable oscillator. In: IEEE international solid-state circuits conference, ISSCC, pp 244–245
18. Nguyen CT-C (2007) MEMS technology for timing and frequency control. In: IEEE transactions on ultrasonics, ferroelectrics and frequency control, pp 251–270
19. Galton I (2002) Delta-sigma data conversion in wireless transceivers. IEEE Trans Microw Theory Tech 50:302–315
20. Wan-Thai Hsu (2006) Reliability of silicon resonator oscillators. In: IEEE international frequency control symposium and exposition, pp 389–392
21. SiT8003XT, SiT8102, and SiT9102 data sheets from SiTime. Available online at: www.sitime.com
22. DSC1018 data sheet from Discera. Available online at: www.discera.com
23. SiTime's product selector sheet. Available online at: http://www.sitime.com/support/product-selector
24. IDT data sheets of MM8102, MM8103, and MM8103. Available online from www.idt.com
25. Hajimiri A (1999) Design issues in CMOS differential LC oscillators. IEEE J Solid-State Circ 34(5):717–724
26. Zannoth M et al (1998) A fully integrated VCO at 2 GHz. IEEE J Solid-State Circ 33(12):1987–1991
27. McCorquodale MS et al (2007) A monolithic and self-referenced RF LC clock generator compliant with USB 2.0. IEEE J Solid-State Circ 42(2):385–399
28. McCorquodale MS et al (2008) A 0.5-to-480 MHz self-referenced CMOS clock generator with 90 ppm total frequency error and spread-spectrum capability. In: IEEE international solid-state circuits conference, ISSCC, pp 350–351
29. McCorquodale MS et al (2008) A 25 MHz All-CMOS reference clock generator for XO-replacement in serial wire interfaces. In: IEEE international symposium on circuits and systems, ISCAS, pp 2837–2840
30. McCorquodale MS et al (2008) Self-referenced, trimmed and compensated RF CMOS harmonic oscillators as monolithic frequency generators. In: IEEE frequency control symposium, pp 408–413
31. McCorquodale MS et al (2009) A 25-MHz self-referenced solid-state frequency source suitable for XO-replacement. IEEE Trans Circ Syst I Regular Pap 56(5):943–956
32. McCorquodale MS et al (2010) A silicon die as a frequency source. In: IEEE international frequency control symposium, pp 103–108
33. Groves R (1997) Temperature dependence of Q and inductance in spiral inductors fabricated in a silicon-germanium/BiCMOS technology. IEEE J Solid-State Circ 32(9):1455–1459
34. McCorquodale MS et al (2011) A history of the development of CMOS oscillators: the dark horse in frequency control. In: IEEE international frequency control symposium, pp 437–442
35. Data sheet of Si500 silicon oscillators. Available from the website of Silicon Labs at: www.silabs.com
36. Wikipedia page on RC oscillators. Available online at: http://en.wikipedia.org/wiki/RC_oscillator
37. McCreary JL (1981) Matching properties, and voltage and temperature dependence of MOS capacitors. IEEE J Solid-State Circ 16(6):608–616
38. St Onge SA et al (1992) Design of precision capacitors for analog applications. IEEE Trans Compon Hybr Manufact Technol 15(6):1064–1071

39. Lane WA et al (1992) The design of thin-film polysilicon resistors for analog IC applications. IEEE Trans Electron Dev 36(4):738–744

40. De Smedt V et al (2009) A 66 μW 86 ppm/°C fully-integrated 6 MHz wienbridge oscillator with a 172 dB phase noise FOM. IEEE J Solid-State Circ 44(7):1990–2001

41. Paavola M et al (2006) A 3 μW, 2 MHz CMOS frequency reference for capacitive sensor applications. In: IEEE international symposium on circuits and systems, ISCAS, pp 4391–4394

42. Blauschild RA (1994) An integrated time reference. In: IEEE international solid-state circuits conference, ISSCC, pp 56–57

43. De Smedt V et al (2009) A 0.4–1.4 V 24 MHz fully integrated 33 μW, 104 ppm/V supply-independent oscillator for RFIDs. In: IEEE European solid-state circuits conference, ESSCIRC, pp 396–399

44. Wikipedia page on wienbridge oscillators. Available online at: http://en.wikipedia.org/wiki/Wien_bridge_oscillator

45. Model 200A Audio Oscillator, 1939, description from HP virtual museum. Available online at: http://www.hp.com/hpinfo/abouthp/histnfacts/museum/earlyinstruments/0002/index.html

46. Hewlett WR (1939) Variable frequency oscillation generator, patent filed on 11 July 1939

47. Johns DA, Martin K (1997) Analog integrated circuits. Wiley, New York

48. Olmos A (2003) A temperature compensated fully trimmable on-chip IC oscillator. In: IEEE symposium on integrated circuits and systems design, pp 181–186

49. Vilas Boas A et al (2004) A temperature compensated digitally trimmable on-chip IC oscillator with low voltage inhibit capability. In: IEEE international symposium on circuits and systems, ISCAS, pp 501–504

50. Choe K et al (2009) A precision relaxation oscillator with a self-clocked offset-cancellation scheme for implantable biomedical SoCs. In: IEEE international solid-state circuits conference, ISSCC, pp 402–403

51. Enz CC et al (1996) Circuit techniques for reducing the effects of op-amp imperfections: autozeroing, correlated double sampling, and chopper stabilization. Proc IEEE 84 (11):1584–1614

52. Gierkink SLJ, van Tuijl E (2002) A coupled sawtooth oscillator combining low jitter with high control linearity. IEEE J Solid-State Circ 37(6):702–710

53. Geraedts et al PFJ (2008) A 90 μW 12 MHz relaxation oscillator with a −162 dB FOM. In: IEEE international solid-state circuits conference, ISSCC, pp 348–349

54. Tokunaga Y et al (2010) An on-chip CMOS relaxation oscillator with voltage averaging feedback. IEEE J Solid-State Circ 45(6):1150–1158

55. McNeill JA (1997) Jitter in ring oscillators. IEEE J Solid-State Circ 32(6):870–879

56. Wikipedia page on ring oscillators. Available online at: http://en.wikipedia.org/wiki/Ring_oscillator

57. Sundaresan K et al (2006) Process and temperature compensation in a 7-MHz CMOS clock oscillator. IEEE J Solid-State Circ 41(2):433–442

58. Shyu Y et al (1999) A process and temperature compensated ring oscillator. In: IEEE Asia Pacific conference on ASICs, pp 283–286

59. Maneatis JG (1996) Low-jitter process-independent DLL and PLL based on self-biased techniques. IEEE J Solid-State Circ 31(11):1723–1732

60. Razavi B (2000) Design of analog CMOS integrated circuits. McGraw-Hill, Boston

61. Pertijs MAP, Huijsing JH (2006) Precision temperature sensors in CMOS technology. Springer, Dordrecht

62. Lee J et al (2009) A 10 MHz 80 μW 67 ppm/°C CMOS reference clock oscillator with a temperature compensated feedback loop in 0.18 μm CMOS. In: IEEE symposium on VLSI circuits, pp 226–227

63. Ueno K et al (2009) A 30-MHz, 90-ppm/°C fully-integrated clock reference generator with frequency-locked loop. In: IEEE European solid-state circuits conference, ESSCIRC, pp 392–395

64. Sebastiano F et al (2009) A low-voltage mobility-based frequency reference for crystal-less ULP radios. IEEE J Solid-State Circ 44(7):2002–2009
65. Sebastiano F et al (2011) A 65-nm CMOS temperature-compensated mobility-based frequency reference for wireless sensor networks. IEEE J Solid-State Circ 46(7):1544–1552
66. Denier U (2010) Analysis and design of an ultralow-power CMOS relaxation oscillator. IEEE Trans Circ Syst-I Regular Pap 57(8):1973–1982
67. Sebastiano F et al (2010) A 1.2-V 10-μW NPN-based temperature sensor in 65-nm CMOS with an inaccuracy of 0.2 °C (3σ) from −70 °C to 125 °C. IEEE J Solid-State Circ 45 (12):2591–2601

Chapter 3
Frequency References Based on the Thermal Properties of Silicon

This chapter introduces the concept of on-chip frequency generation based on the thermal properties of silicon. Thermal-diffusivity of silicon, D, will be introduced as the rate at which heat diffuses through a silicon substrate. It will be described how an electrothermal filter (ETF) harnesses this physical property in order to produce accurate on-chip delays. The design of a practical ETF within standard CMOS processes will be described. It will be shown that an ETF behaves like a low-pass filter with a defined phase shift, which is a combined function of its geometry and D. Furthermore, a method of frequency generation based on an electrothermal frequency-locked loop (FLL) will be introduced. Such loop locks the output frequency of a variable oscillator to the phase shift of an ETF.

The accuracy of the output frequency of an electrothermal FLL is a function of its ETF. The latter is mainly determined by geometry, whose accuracy is determined by lithography. As predicted by Moore's law, the minimum feature size of IC technologies shrink, which implies that their lithographic accuracy improves. This means that the accuracy of an electrothermal FLL should benefit from process scaling, and so, should be a good basis for an electrothermal frequency reference. Furthermore, this chapter discusses the dynamics of an FLL, as well as the effect of the ETF's thermal noise on the output jitter of the loop.

3.1 Introduction

Previous chapter provided an overview of on-chip frequency references. This survey shows that thermal frequency references have hardly been investigated. Some applications of electrothermal [1–3] systems have been investigated previously. For instance, studies have been conducted on the realization of very low cut-off frequency filters [4], high-Q band-pass filters [5], and thermal oscillators [6–8].

S.M. Kashmiri and K.A.A. Makinwa, *Electrothermal Frequency References in Standard CMOS*, Analog Circuits and Signal Processing, DOI 10.1007/978-1-4614-6473-0_3, © Springer Science+Business Media New York 2013

Fig. 3.1 The conceptual inclusion of a thermal-delay line in an electrothermal system, i.e. a thermal oscillator

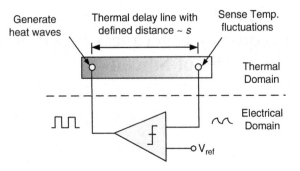

The heat generated at a given point in a silicon chip diffuses through its substrate and can be sensed some distance s away. The time it takes for heat to travel between the heater and the temperature sensing points is determined by the thermal-diffusivity of silicon. Early studies on silicon substrates show that thermal-diffusivity, D, of IC-grade silicon is a well-defined parameter [9, 10]. The stability of D, when combined with accurate geometries, makes it possible to realize accurate thermal delays.

The ability to realize an accurate on-chip delay means that an accurate period of oscillation, and hence an accurate frequency, can be defined. The application of this concept is illustrated by the thermal feedback oscillator shown in Fig. 3.1. Here, the digital output of a comparator drives a heating element in a silicon substrate. The heat waves diffuse through the silicon and cause temperature fluctuations at a distance s from the heat source, where a temperature sensor converts them back into an electrical signal. The delay between this signal and the power dissipation is the thermal delay. The oscillator's period of oscillation will be thus determined by s and D.

The thermal delay line described above can be regarded as an electrothermal filter (ETF). Due to the thermal inertia of the substrate, ETFs behave like low pass filters. As will be shown later in this chapter, an ETF can be implemented as a standard CMOS element. This will be the heart of the electrothermal frequency references described in the rest of the book. After a brief review of the thermal properties of silicon, the modeling and design of ETFs will be discussed. This will then be followed by a description of the use of ETFs in the generation of accurate frequencies.

3.2   Thermal Properties of Silicon

To provide a general understanding of the electrothermal systems discussed throughout this book, it is useful to review the thermal properties of materials with a special focus on silicon. This is especially useful for modeling ETFs. The phenomenon of interest is the transfer of heat, which is physically possible through

Table 3.1 Analogy between the thermal and electrical domain quantities

Thermal quantity	Electrical quantity
Temperature T (Kelvin)	Voltage (V)
Heat flow P (Watt)	Current I (A)
Heat Q (Joule)	Charge Q (C)
Thermal resistance R_{TH} (Kelvin/Watt)	Resistance R (V/A)
Thermal conductance k (Watt/milli-Kelvin)	Conductance (A/V)
Heat capacitance C (Joule/Kelvin)	Capacitance C (Q/V)

three major mechanisms. These are *conduction*, in which heat diffuses through a material, *convection*, in which the flow of a gas or a fluid transfers heat, and *radiation*, in which energy is transferred through the emission of electromagnetic waves. Since microelectronic structures are solid bodies of material with relatively small geometries, the *conduction* of heat is the most dominant mechanism of heat transfer [11].

There is an analogy between thermal and electrical systems. In a thermal system, an amount of heat denoted by Q (Joules) flows to produce a heat flow P (Watts), such that $P = Q/t$, where t is the time. This resembles the electric charge Q (Coulombs) flowing through a conductor and leading to a current I (Amperes). A heat flux q (a heat flow Q per area A) is physically induced by a temperature gradient ΔT (Kelvin) such that [11]:

$$q = -k\nabla T. \tag{3.1}$$

This is Fourier's law, in which factor k is the *thermal conductivity* of the material in W/mK. This implies that temperature is analogous to electrical voltage, because an electrical voltage applied to a conductance (resistance) induces a current flow in the material. The negative sign in (3.1) implies that heat flows from hot to cold areas. Based on this equation, a thermal resistance R_{TH} can be defined. The thermal resistance of a material is the temperature difference ΔT across it, divided by the heat flow Q, such that:

$$R_{TH} = \frac{\Delta T}{Q}. \tag{3.2}$$

Using the same methodology, a thermal capacitance can also be defined as an analogous parameter to the electrical capacitance. An amount of heat (in Joules) applied to a thermal capacitance C (Joule/Kelvin) leads to a temperature increase that is proportional to C. Table 3.1 summarizes the analogies of the thermal and electrical parameters.

The concept of thermal resistance R_{TH} can be extended to the concept of thermal impedance Z_{TH} in the case when AC heat flow and AC temperature fluctuations are considered. This will be the case in an electrothermal filter (ETF). The reason for this is that the power $Q(\omega)$ dissipated in the heater is an AC parameter, which is

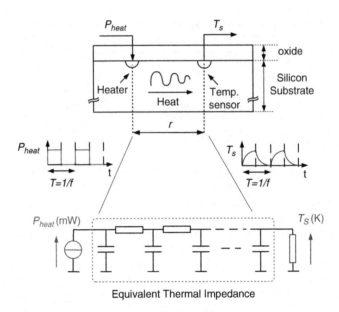

Fig. 3.2 A simplified thermal delay line made of a point-heater and point-temperature sensor located at the surface of a silicon chip, and an electrical domain equivalent to that made with an RC network

caused by the electrical signal applied to the heater. This AC power dissipation leads to temperature fluctuations $\Delta T(\omega)$. A frequency dependent thermal impedance $Z_{TH}(\omega)$ can then be expressed as:

$$Z_{TH}(\omega) = \frac{\Delta T(\omega)}{Q(\omega)}. \qquad (3.3)$$

To obtain intuition by means of a simplified model [3, 12], an ETF can be reduced to the combination of a point-heater and a point-temperature sensor. These are located at a distance r apart from each other on the surface of a silicon substrate (see Fig. 3.2). Once this structure can be modeled, the results can be extended to the more complex geometries of actual electrothermal filters. Figure 3.2 also shows the passivation oxide layer that covers the silicon substrate; however, this may be initially regarded as being an ideal heat insulator. This is because the thermal conductivity of silicon oxide is nearly two orders of magnitude less than that of bulk silicon [11].

The lower half of Fig. 3.2 shows an analogy between the thermal structure and an electrical domain circuit. The electrical circuit consists of a network of RC electrical elements. The current source at the left represents the AC power dissipation P_{heat} in the heater. The temperature fluctuations, denoted by T_S at the temperature sensor side, can be modeled by a voltage across a load impedance. Such an electrical model of an ETF is useful especially for time-domain behavioral modeling. This will be introduced further through the chapter. To study the behavior of ETFs in the frequency-domain, an analytical thermal model is more convenient. This will be discussed next.

To investigate the behavior of the point-heater point-sensor structure shown in Fig. 3.2, its thermal impedance can be calculated in the frequency domain. This is defined as the transfer function that relates the AC temperature rise $T(\omega)$ at the temperature sensor side of the thermal delay line to the power $P(\omega)$ dissipated in the heater (see Fig. 3.2). In order to derive this characteristic the linearized heat diffusion equation should be solved in the frequency domain [12]:

$$\nabla^2 T(\omega, r) - \frac{j\omega}{D} T(\omega, r) = 0. \tag{3.4}$$

where j is the complex operator and ω is the angular frequency. The parameter D (in $cm^2 s^{-1}$) is the thermal diffusivity of the heat conducting medium, which, in the structure of Fig. 3.2, is the thermal diffusivity of silicon. This is defined as:

$$D = \frac{k}{\rho c_p}. \tag{3.5}$$

with k thermal conductivity of silicon, ρ its density and c_p its specific heat. The latter relates the temperature rise of a unit volume of silicon to the amount of absorbed heat. It should be noted that D has a temperature-dependent behavior, with a room-temperature value of 0.88 $cm^2\ s^{-1}$ reported in the literature (see [22], Chap. 1). Its temperature dependence over the industrial temperature range can be approximated with a power law [10, 13]:

$$D \propto \frac{1}{T^n}. \tag{3.6}$$

Based on the recommended values of D for high-purity silicon over a temperature range of 250 K and 400 K, n is approximately 1.8 [10]. Furthermore, thermal diffusivity is not affected by doping levels normally used in the mono-crystalline IC-grade silicon substrates. This is because its deviation from the values measured for the pure silicon is noticeable only for doping levels $>10^{17}$ cm^{-3} [9].

The thermal impedance $Z_{TH}(\omega)$ can then be calculated by solving (3.4), considering a boundary condition at $\omega = 0$, as defined by (3.1) [12]:

$$\frac{Q}{A(r)} = -k \frac{\partial T(\omega, r)}{\partial r} \bigg|_{\omega=0}. \tag{3.7}$$

where A is the area. The effect of the oxide can be considered as a perfect insulator that reflects the heat flux downwards [12]. Therefore, the system is symmetrical in a sphere around the heater and thus the temperature around it will be a function of

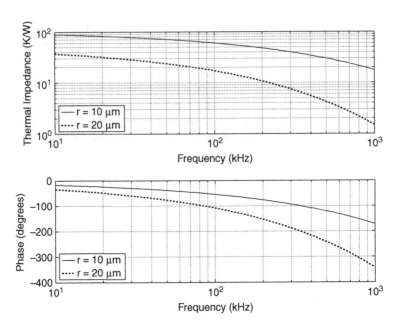

Fig. 3.3 Thermal impedance at various distances of r from a point-heater source over frequency

distance r. In this manner (3.4) can be solved, and then the thermal impedance of the structure can be calculated as [13]:

$$Z_{TH}(\omega, r) = \frac{T(\omega, r)}{P(\omega)} = \frac{1}{2\pi k r} \exp\left(-r\sqrt{\frac{\omega}{2D}}\right) \cdot \exp\left(-jr\sqrt{\frac{\omega}{2D}}\right). \qquad (3.8)$$

This thermal impedance has a magnitude and a phase response of:

$$\begin{aligned} &Magnitude : |Z_{TH}| = \tfrac{1}{2\pi k r} \exp\left(-r\sqrt{\tfrac{\omega}{2D}}\right) \\ &Phase : \arg(Z_{TH}) = \phi_{TH} = -r\sqrt{\tfrac{\omega}{2D}} \end{aligned} \qquad (3.9)$$

The phase shift ϕ_{TH} of the thermal impedance, is a function of r and ω and has a low-pass characteristic. Figure 3.3 shows the magnitude and phase of the thermal impedance as a function of frequency for two different structures with $r = 10\ \mu m$ and $20\ \mu m$. It shows that the longer the thermal delay line, the greater is the phase shift and the attenuation it introduces to the temperature fluctuations at the sensor. The magnitude of the thermal impedance gives an estimate of the expected thermal signal amplitude for certain levels of heater power dissipation. For instance when $r = 20\ \mu m$, 1 mW of power dissipation in the heater results in less than 20 mK temperature variations at the temperature sensor. This low signal level reflects that silicon is a good conductor of heat and also that a lot of the dissipated heat will be lost to the substrate.

If the purity of the silicon substrate is such that the value of D is stable over process, the accuracy of ϕ_{TH} will be determined by the accuracy of the distance r. In a CMOS process this is defined by the accuracy of the lithography. In the case of the point-heater point-sensor model of Fig. 3.2, lithographic inaccuracy induces an error in r, which leads to an error in the phase shift of the structure. When an oscillator is locked to an ETF's thermal phase shift (like in Fig. 3.1), any variations in this phase causes frequency errors. Since D is temperature-dependent, if (3.6) and (3.9) are combined, the effect of variations in r on the variations of ϕ_{TH} can be calculated as:

$$\frac{d\phi_{TH}}{\phi_{TH}} = \frac{dr}{r} = \frac{n}{2} \cdot \frac{dT}{T}. \tag{3.10}$$

this shows that for larger values of r, the effect of lithographic spread on ϕ_{TH} is smaller. Furthermore, this shows that the lithographic-induced phase error increases linearly with temperature.

The thermal impedance phase relation can be further used to study the phase-frequency sensitivity of an ETF for a given temperature. For a fixed r and T (thus fixed D), differentiating (3.9) with regard to frequency results in:

$$\frac{d\phi_{TH}}{\phi_{TH}} = \frac{1}{2} \cdot \frac{d\omega}{\omega} = \frac{1}{2} \cdot \frac{df}{f}. \tag{3.11}$$

which shows how relative variations in ETF phase, e.g. due to lithographic errors, can be expressed in terms of variations in its drive frequency.

3.3 Electrothermal Filters in CMOS

An ETF can be realized by integrating a heater in close proximity to a relative temperature sensor. In CMOS technology, any kind of diffusion resistor can be used as a heater, i.e. any resistor located in the substrate. In this case, silicon will be the major heat conducting medium. Transistors can also be used as heaters.

The relative temperature sensor can be implemented by various devices: resistors, transistors, or thermopiles. The drawback of using integrated resistors as temperature sensors is their rather low temperature coefficients (in the order of a few thousand ppm/K). Furthermore, they require biasing, which can lead to extra power dissipation and self-heating. Other possible temperature sensors are bipolar transistors. They have a relatively strong temperature dependence (about 2 mV/K), but they also need biasing.

A better and simpler solution is to use thermopiles as relative temperature sensors. Thermopiles are made of series-connected thermocouples. A thermocouple consists of the ohmic connection of two different conducting materials (see Fig. 3.4).

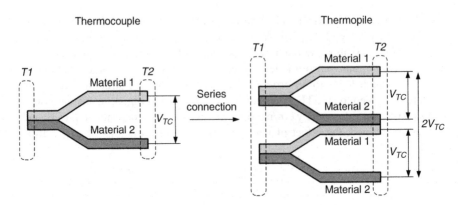

Fig. 3.4 A thermocouple and the series connection of thermocouples

When a temperature gradient is applied across the structure a voltage V_{TC} is induced between the other terminals: the *Seebeck* effect [14]. This voltage is proportional to the temperature difference $(T_1 - T_2)$ due to the gradient and to the combined Seebeck coefficient of the two materials. As shown by Fig. 3.4, the series connection of thermocouples adds up their *Seebeck* voltages like a battery, forming a thermopile. It should be noted that a thermopile has two junctions. When used in an ETF, the junction closest to the heater (the $T1$ side in Fig. 3.4) is called the *hot junction* and the one furthest away from the heater (the $T2$ side in Fig. 3.4) is called the *cold junction*.

Thermocouples can be realized simply in any standard CMOS process, by exploiting the contact between a p^+ or n^+ diffusion and the Aluminum or Copper interconnect. This results in a sensitivity of about 0.5 mV/K [13]. Compared to other types of relative temperature sensors, thermopiles do not require biasing, and are free of offset. Furthermore, it should be noted that a thermopile can be realized in an n-well. This helps shielding it from the substrate noise. In this case, the thermopile can be realized only with p^+ diffusion material.

Another advantage of thermopiles is that the temperature sensing occurs at the junction between the two different materials, which can be made very small. In the case of a p^+/Aluminum thermocouple, the junction is formed by a contact hole between the Aluminum and the diffusion, which can have near-minimum dimensions. This can be compared conceptually to the area of a minimum-size bipolar transistor in a CMOS process shown in Fig. 3.5. The contact hole forming the thermocouple's *hot* or *cold* junctions is almost the same size as the transistor's emitter contact, while the temperature sensing base-emitter junction is much larger. This large area introduces uncertainty in the exact distance between the heater and the temperature sensing point [8], affecting the accuracy of ETF.

Based on the previous discussions, it can be concluded that thermopiles are the best temperature sensors for ETFs due to their simplicity and precise geometry, the lack of biasing, and the possibility of increasing their output signal by stacking thermocouples. Since thermopiles are the selected temperature sensor structures for ETFs, we need to revisit the simplified point-heater point-sensor model shown

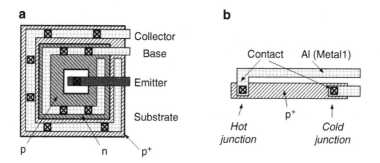

Fig. 3.5 Conceptual comparison of the area of a substrate PNP transistor (**a**), with contact holes forming the junctions of a thermocouple (**b**)

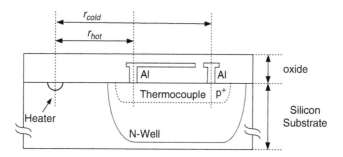

Fig. 3.6 Side view of a point-heater and thermocouple structure

in Fig. 3.2. Since thermopiles measure temperature differences, the thermal impedance seen by a thermopile will now be affected by the location of its cold junction. This can be seen in the new model shown in Fig. 3.6. Because the AC thermal impedances are frequency domain phasors, the thermal impedance of a thermocouple is the difference between the thermal impedance of the *hot junction* and the thermal impedance seen at the *cold junction*:

$$Z_{THP}(\omega) = Z_{TH}(\omega, r_{hot}) - Z_{TH}(\omega, r_{cold}). \qquad (3.12)$$

To increase the efficiency of an ETF, a differential structure can be used (see Fig. 3.7). This also helps reduce the effect of variations in the position of the heater relative to the position of the thermocouple junctions. If the heater is shifted away from one thermocouple by Δr (see Fig. 3.7) it will be closer to the opposite thermocouple by the same distance Δr. If Δr is small compared to r, it can be assumed that the phase shift of the thermal impedance between the heater and a hot junction is a linear function of r, and so the overall phase response of the differential structure should not be altered by this error.

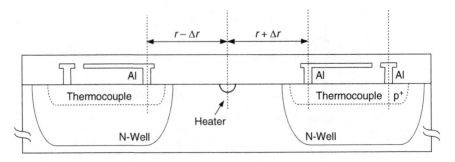

Fig. 3.7 Side view of a differential point-heater and thermocouple structure

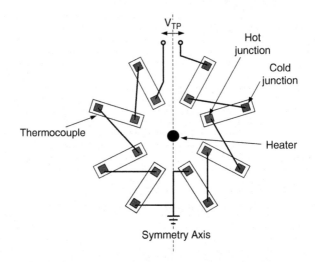

Fig. 3.8 Top view of a conceptual differential ETF structure with a differential thermopile that is symmetrically located around its point heater

The top view of a conceptual differential ETF is shown in Fig. 3.8. As shown here, multiple thermocouples could surround the heater at both sides. These could be arranged with their *hot* and *cold* junctions located on equi-center circles. A differential voltage V_{TP} is then available at the outputs of this ETF.

3.4   ETF Design

The design of a practical ETF involves the choice of appropriate heater and thermopile geometries. This will be discussed in the following subsections, which will then be followed by two design examples of ETFs in a 0.7 μm CMOS process.

3.4.1   General Heater Considerations

A practical ETF does not have a point heater, since a practical heater will have a certain width and length. Therefore, in order to design a real ETF we need to be able to derive its thermal impedance considering the heater geometry. This makes the analysis rather complex.

To simplify matters, the heater may be sub-divided into several unit cubes [12, 15]. Since the width of each cube is much smaller than the distance to the thermocouple junctions, each cube can be modeled as a point-heat source. This way, the equivalent thermal impedance of each cube to all the hot and cold junctions of the thermopile can be calculated. As a result, the thermal impedance of the whole ETF can be derived by adding up the vector sum of all the calculated thermal impedances.

For a total number of M divided cubes, and the number of *hot* and *cold* thermopile junctions equal to P_{hot} and P_{cold}, the total thermal impedance of a practical structure can be calculated analytically from:

$$Z_{TH,ETF} = \frac{1}{M} \sum_{m=1}^{M} \sum_{p=1}^{Phot} Z_{TH}(\omega, r_{m,p}) - \frac{1}{M} \sum_{m=1}^{M} \sum_{q=1}^{Pcold} Z_{TH}(\omega, r_{m,q}). \qquad (3.13)$$

This implies that further calculations on ETF characteristics require numerical vector-based calculations that can best be done by mathematical calculation software such as Matlab.

3.4.2   General Thermopile Considerations

Design of a thermopile involves choices regarding the total number of thermocouple arms and their lengths. These will be discussed in the following paragraphs.

The more thermocouples in series, the larger the output signal of the ETF, but one should consider that each extra thermocouple arm contributes extra thermal noise (caused by extra resistance of each arm). The signal-to-noise-ratio (SNR) at the output of the thermopile determines the resolution with which the ETF's thermal phase shift can be measured. Considering n thermocouples each with a resistance R_{tc}, the SNR can be calculated within a bandwidth of $BW = 1$ Hz:

$$SNR = \frac{P_{signal}}{P_{noise}} = \frac{[n \cdot S_{tp} \cdot \Delta T]^2}{n \cdot 4kTR_{tc}} = n \cdot \frac{[S_{tp} \cdot \Delta T]^2}{4kTR_{tc}}. \qquad (3.14)$$

where S_{tp} is the Seebeck coefficient of the Aluminum and p^+ contact, ΔT is the temperature difference between the *hot* and *cold* junctions, T is the ambient temperature, and k is Boltzmann constant. Since for a given heater geometry, the thermopile area will be fixed, and so, increasing n, increases the resistance of thermopile, and hence the SNR remains constant. However, increasing n results in a larger output signal and thus the number of thermocouples should be maximized as much as possible. However, each p^+ arm of the thermopile exhibits a parasitic junction capacitance, and thus the ETF also exhibits electrical filtering. Therefore, a thermopile behaves like a distributed RC filter, which introduces electrical phase shift. Increasing n will increase this parasitic electrical phase shift. Consequently, the optimal value of n involves a trade-off between the signal level and electrical phase shift.

For a fixed number of thermocouples, increasing the length of a thermocouple means placing the *cold* junctions further away from the *hot* junctions. This means that, on the one hand, the increase in the resistance of the thermopile due to the increase in its length increases its thermal noise, while on the other hand, setting the *cold* junctions further away increases the temperature difference between the two junctions. This is because the thermal impedance due to the *cold* junction drops, and therefore the total thermal impedance increases (see (3.9) and (3.13)). This increase in the temperature difference will again increase the signal at the output of the thermocouple, implying that an optimum value of the thermocouple length for achieving a maximum SNR should exist. This value can be found numerically through an optimization algorithm.

This algorithm can be shown for the point-source system (see Fig. 3.6), by considering the *hot* and *cold* junction distances r_h and r_c from the heater, respectively. For a p^+ diffusion sheet resistance $R_\square$, and thermocouple width W, the thermocouple noise power in a bandwidth of $BW = 1$ Hz will be:

$$P_{noise} = 4kT \frac{L}{W} R_\square.$$ (3.15)

where $L = r_c - r_h$, is the thermocouple length. The signal power at the output of one thermocouple is:

$$P_{signal} = \left[S_{tp} \cdot \Delta T \right]^2.$$ (3.16)

where ΔT is the temperature difference between the cold and the hot junction. This is a function of the thermal impedance of the point-source structure Z_{THP}, and the heater power dissipation P_{heater}. The AC temperature variation ΔT can be calculated from (3.12) and (3.8) by:

$$\begin{aligned} \Delta T(\omega) &= P_{heater}(\omega) \cdot Z_{THP}(\omega) \\ &= P_{heater}(\omega) \cdot [Z_{TH}(\omega, r_h) - Z_{TH}(\omega, r_c)] \end{aligned}.$$ (3.17)

which can be written as:

$$\Delta T(\omega) = P_{heater}(\omega) \cdot [Z_{TH}(\omega, L - r_c) - Z_{TH}(\omega, r_c)].$$ (3.18)

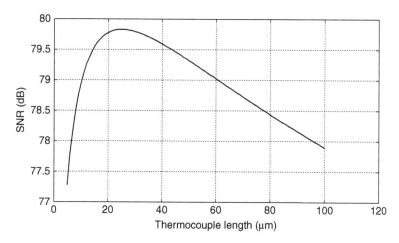

Fig. 3.9   A point-source structure's SNR as a function of thermocouple length L

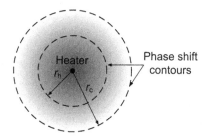

Fig. 3.10   Equi-phase shift contours from a point heat source

As an example we consider a fixed $r_h = 25\mu m$, $\omega = 120$ kHz, $K_{seebeck} = 0.5$ mV/K, $P_{heater} = 10mW$, $R_\square = 200$ Ω/square, $W = 1\mu m$, and solving (3.15, 3.16, 3.17, and 3.18) numerically at room temperature and the bandwidth $BW = 1$ Hz for 5 $\mu m \leq L \leq 100$ μm, the SNR of the structure can be calculated in dB (see Fig. 3.9). This figure shows that for this specific example, the optimum thermocouple length is 25 μm, for which the structure's SNR is maximized. Further increasing L increases the signal by reducing the cold junction's thermal impedance, but it also increases the thermal noise, which causes a loss of SNR.

In an ETF, the heat generated by a point-heat source diffuses uniformly in all directions. Therefore, all points located at the same distance from the heater will have equal thermal impedances and thus equal phase shifts (see Fig. 3.10). This translates into circles with a common center point at the heater. Locating all the *hot* junctions of a thermopile on a circle with radius r_h, and all the *cold* junctions on another circle with radius r_c, ideally translates into thermocouple signals that have the same phase. Their signals, when added together, cause the maximum output signal. These circles are called equi-phase contours. In practice, when the layout

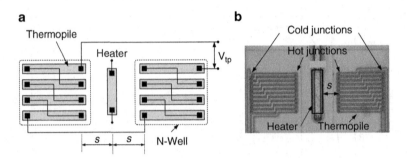

Fig. 3.11 (**a**) Simplified layout, and (**b**) photomicrograph of an earlier ETF with a bar heater (bar ETF)

of an ETF is drawn within the design rule constraints of a CMOS process, it is difficult to draw the p^+ arms arranged on the spikes originating from the center of the circles (like Fig. 3.8). Therefore, to simplify the layout, horizontal orientation of the thermocouple arms can be kept. The hot junctions can be fixed on the circle, and the location of the cold junctions can be calculated numerically for maximum SNR [15].

3.4.3   Design of a Bar ETF

An ETF designed in a 0.7 μm CMOS process with bar-shaped heater and rectangular thermopile [13] is shown in Fig. 3.11. The distance between the heater and the thermopile, denoted by s is 20 μm. For layout simplicity the thermocouple arms all have the same length and the *hot* and the *cold* junctions of the thermopile are arranged in parallel vertical lines, such that the thermocouples could be placed horizontally (the layout of the thermocouple p^+ arms is thus much simpler than the conceptual layout shown in Fig. 3.8). Numerical simulations of the thermal impedance of the bar ETF have been performed using (3.13) and the results are shown in Fig. 3.12.

3.4.4   Design of an Optimized ETF

With the thermopile optimization considerations in mind, another ETF [16], which is optimized compared to the bar version shown in Fig. 3.11 can be designed. A simplified layout and a photomicrograph of the optimized ETF are illustrated in Fig. 3.13. Here, the layout has been optimized to maximize the SNR at the output of the thermopile, while maintaining the same phase-shift of the previous ETF (about 90°) at about 100 kHz. The amplitude and phase response of the thermal impedance

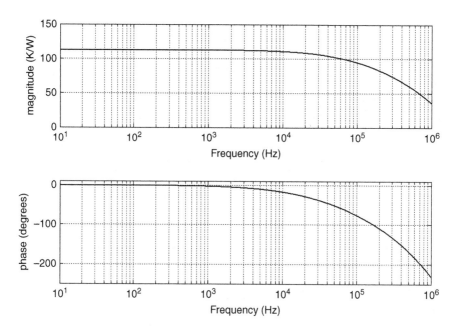

Fig. 3.12 Amplitude and phase response of the thermal impedance associated with the bar ETF at room temperature

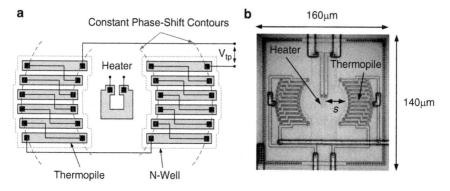

Fig. 3.13 (**a**) Simplified layout, and (**b**) photomicrograph of an optimized ETF

associated with this structure have been numerically modeled in MATLAB and the results are illustrated in Fig. 3.14.

For the same power dissipation, reducing the heater area leads to a greater concentration of heat [15], and, thus, to larger temperature variations in the substrate. For this reason, the optimized ETF's heater has a smaller area compared to the bar ETF. To achieve the same order of heater resistance (1.2 kΩ), the resistor is folded compared to the previous rectangular heater [15]. In addition, the

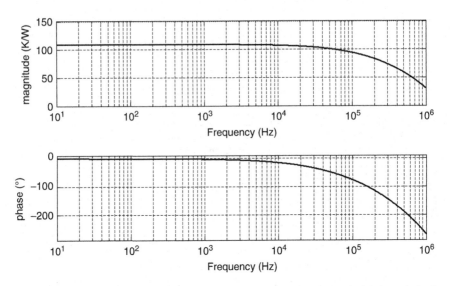

Fig. 3.14 Amplitude and phase response of the thermal impedance associated with the optimized ETF at room temperature

thermopile's *hot* junctions are located on a roughly circular, equi-phase contour, which maximizes its output amplitude. Finally, the length of each arm, and hence its thermal noise contribution, is chosen to maximize the SNR at the thermopile's output. Compared to the earlier ETF [13], the number of thermocouples are increased from 20 to 24. The thermopile resistance is reduced from 36 to 20 kΩ, leading to a thermal noise reduction from 24 to 18 nV/$\sqrt{Hz}$. For the same heater power dissipation, these changes lead to a 50% SNR increase.

The thermopile of the optimized ETF has a total resistance of 20 kΩ and a total capacitance of 600 fF. Simulations show that at an excitation frequency of 85 kHz, the resulting electrical phase shift due to the electrical filter formed by these values is only 0.14°, which is much smaller than the thermal phase shift of the ETF. However, the electrical phase shift will spread by tens of percent over process corners due to spread in the absolute doping levels of the p$^+$ arms and the n-well. This forms another bound on the absolute accuracy of ETFs.

3.5 Modeling for Time-Domain Analysis

With the knowledge built so far, a prediction of the frequency response of a given ETF structure can be derived from its thermal impedance. However, a model based on a network of RC elements such as the one shown in Fig. 3.2 is more useful for time-domain simulations of the structure. An equivalent Foster form [17] of the ETF thermal impedance can be made such that its output temperature variation ΔT is a function of the power P dissipated in the heater [17]. This network is shown in

Fig. 3.15 Electrical-domain equivalent model of an ETF's thermal impedance

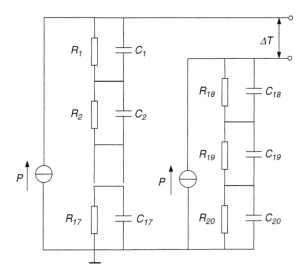

Table 3.2 The approximate RC section values of the Foster network

Resistance	R1	R2	R3	R4	R5	R6	R7	R8	R9	R10
Value	0.59	3.35	1.25	0.65	4.81	10.57	12.59	9.59	5.86	5.65
Capacitance	C1	C2	C3	C4	C5	C6	C7	C8	C9	C10
Value	9.6E-8	4.4E-8	1.7E-7	1.5E-6	3.3E-7	2.3E-7	3.1E-7	6.9E-7	1.7E-6	2.9E-6
Resistance	R11	R12	R13	R14	R15	R16	R17	R18	R19	R20
Value	4.99	3.35	2.09	1.06	0.78	0.41	0.41	1.97	8.61	2.95
Capacitance	C11	C12	C13	C14	C15	C16	C17	C18	C19	C20
Value	5.2E-6	1.3E-5	3.2E-5	1.1E-4	2.2E-4	7.1E-4	18E-4	4.6E-8	4.1E-8	1.8E-7

Fig. 3.15, including 20 RC sections whose element values are summarized in Table 3.2. Using this network, the frequency response of an ETF predicted by the thermal impedance model [11] (magnitude and phase response) as well as the structure's step response could be reproduced with a reasonable accuracy (see Fig. 3.16). The above-mentioned RC network has been used in the time-domain simulations presented throughout this book. More details about this method of time-domain modeling are described in Appendix A.

3.6   Thermal Oscillators

The feasibility of on-chip frequency generation based on the thermal properties of silicon was initially shown by the thermal oscillators described in [6, 7]. These were relaxation oscillators whose frequencies were determined by a thermal delay. The target application of the oscillators in [6, 7] was to monitor the temperature of microelectronic structures.

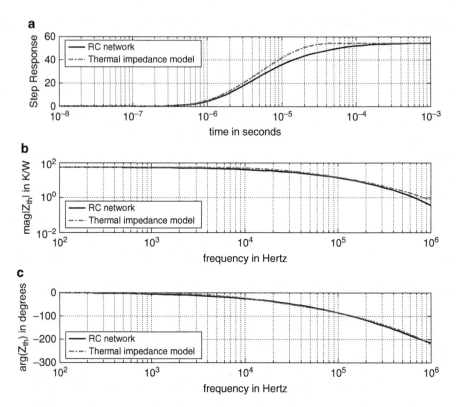

Fig. 3.16 Comparison of the thermal impedance model and the Foster RC network model prediction for an ETF: (**a**) step response, (**b**) magnitude, and (**c**) phase response

A block diagram of the first thermal relaxation oscillator is shown in Fig. 3.17. The output of the thermal delay line is fed back to its heater via an amplifier. A gain-control block maintains unity loop-gain and an offset-control block regulates the common-mode level of the oscillating wave referred to ground. The gain-control block and the amplifier monitor the signal from the temperature sensor. As mentioned earlier in this chapter, silicon is a good conductor of heat and therefore the thermal delay line is rather inefficient. This means that even with large levels of power dissipation in the heater (10–20 mW) the thermal delay line's output would only have been in the order of a few milli-volts. This small signal is accompanied by the wide-band thermal noise generated by the temperature sensor. The result is excessive jitter at the output frequency. For the thermal oscillator in [7], the jitter amounted to tens of percent of an oscillation period. The output frequency varied from 91 to 74 kHz as the temperature varied from 20°C to 70°C, respectively.

A thermal-feedback oscillator incorporating thermocouples was proposed by Bosch in [8]. This oscillator used on-chip thermocouples realized by "p" or "n" type semiconductor materials in contact with Aluminum. Bosch located a heater ~200 μm apart from the thermocouple (see Fig. 3.18). An amplifier was then

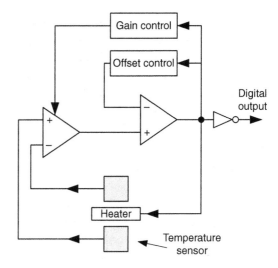

Fig. 3.17 A block diagram of the first relaxation thermal feedback oscillator

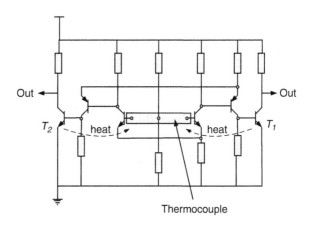

Fig. 3.18 Thermal feedback oscillator of Bosch incorporating an on-chip thermocouple

feeding back the output of the thermocouple to heat-dissipating transistors T_1 and T_2 (bipolar devices used as heaters), forming the complete thermal feedback oscillator. Bosh reported a nominal oscillation frequency of 200 kHz. One of the main motivations of this work described in [8] was to fill up the lower frequency range below 300 kHz, which was not covered by RC oscillators at that time.

Another thermal feedback oscillator implemented in a 2-poly 0.35 μm CMOS process is reported in [18]. This is a phase-shift oscillator that uses an inverting amplifier and an ETF in its feedback path. The ETF was built by means of polysilicon heaters and MOS transistors operated in the sub-threshold region as temperature sensors (see Fig. 3.19). The silicon oxide between the heater and the MOS transistor forms the main thermal path.

In [18] a phase shifting amplifier was made by combining an electrothermal filter and a differential amplifier (see Figs. 3.19 and 3.20). A heater driving circuit drives

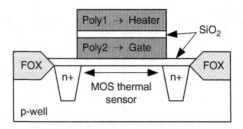

Fig. 3.19 The electrothermal filter made by the 2-poly process with a poly heater and a MOS as the temperature sensor

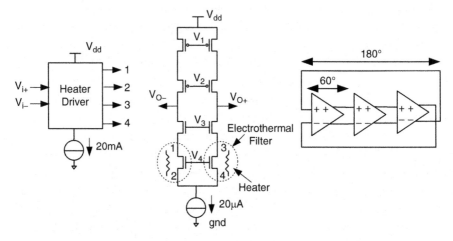

Fig. 3.20 Heat conducting amplifier made of a heater drive and a differential amplifier combined with an electrothermal filter (*left* and *center*), and a triple-phase oscillator incorporating three of the heat-conducting amplifiers

the heater and the phase shifted signal picked up by the electrothermal filter's MOS transistors is amplified. At the oscillation frequency, the output of the phase shifting amplifier V_{out} will have a phase shift of 60°. By cascading three of these amplifiers in a loop and feeding back the output of the third amplifier to the first one, a total loop phase of 180° was made. The structure is able to oscillate at a frequency of 1.25 MHz. The total current through the three heaters of its electrothermal filters amounts to 60 mA, which assuming a 3.3 V supply of the 0.35 μm process, translates to about 60 mW. No jitter or noise measurement results are reported in [18].

3.7 Electrothermal Frequency-Locked Loops

The work published in [6, 7], showing the feasibility of a thermal oscillator, was followed by an improved architecture based on an electrothermal frequency-locked loop (FLL) [19]. The loop incorporates a thermal filter, made of two heaters and a

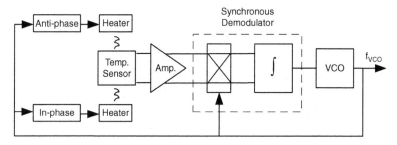

Fig. 3.21 Block-diagram of an earlier thermal frequency-locked loop

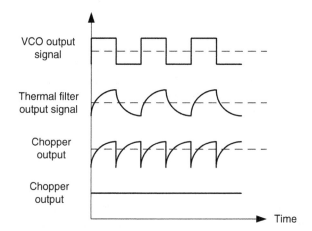

Fig. 3.22 Timing-diagram of an electrothermal frequency-locked loop

thermopile, together with a VCO (Fig. 3.21). The heaters of the ETF are driven in anti-phase by the VCO, creating an AC temperature variation polarity across the thermopile. The feedback loop ensures that the VCO operates at a frequency that corresponds to a fixed phase shift in the thermal filter. Compared to the other thermal oscillators [6, 18], the key element introduced in [19] was the use of a *synchronous demodulator* as a phase detector and narrow-band tracking filter. This narrow-band filtering limits the loop's noise bandwidth to a few Hertz, resulting in low VCO jitter at reasonable heater power dissipation.

In the system shown in Fig. 3.21 the VCO drives the thermal filter with a square-wave signal. The resulting low-pass filtered output signal is amplified and applied to a chopper, which periodically reverses the polarity of the signal. Since the chopper is driven with the same signal that excites the thermal filter's heater, it operates as a synchronous demodulator. This means that its average output will be a function of the thermal filter's phase shift. The presence of an integrator in the forward path of the loop forces the VCO to oscillate at a frequency where the average output of the chopper is zero. This corresponds to a thermal phase shift of 90° (Fig. 3.22) and hence the frequency corresponding to this phase shift is called f_{90}.

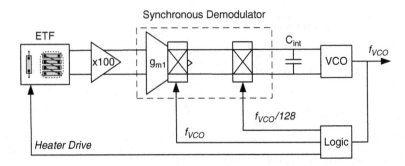

Fig. 3.23 Block-diagram of an electrothermal FLL with low-frequency chopping

The synchronous detection and integration gives the implementation of [19] an important advantage over [6, 7, 18]. This advantage is mainly the insensitivity of its output frequency to the thermal filter's output signal amplitude, which can vary due to process spread. Also, the narrow band filtering provided by the synchronous detector behaves like a narrow-band tracking filter centered around the f_{VCO}, reducing the effect of the thermal filter's thermal noise on the output jitter.

In [19], the thermal filter's readout amplifier had a gain of 100 and a bandwidth of 2.5 MHz. A heater power dissipation of 6.5 mW resulted in a thermal filter output signal of 500 μV_{p-p} (reported distance from heater to heater is 100 μm and the heater distance to each side of the temperature sensor is less than 10 μm). In a loop bandwidth of 1 Hz and at room temperature, an f_{VCO} of 50 kHz and a jitter of 0.1% were reported. The device-to-device frequency spread (eight devices) measured from $-50°C$ to $125°C$ was $\pm 1\%$ and showed a temperature dependence of $T^{-1.3}$, where T is the absolute temperature.

Followed by the work of [19], an improved temperature-to-frequency converter based on an electrothermal FLL was proposed in [13]. Compared to prior art, this work used a readout amplifier with a larger bandwidth and lower offset. These resulted in less device-to-device spread of the output frequency, which led to more accurate temperature sensing. Furthermore, the temperature dependence of the loop was locked more accurately to the thermal-diffusivity of silicon. The ETF used in [13] is the bar ETF shown in Fig. 3.11.

A block diagram of the [13] FLL is shown in Fig. 3.23. A pre-amplifier amplifies the output signal of the thermopile before it is processed by a transconductor g_m. The transconductor's output current is then multiplied by a chopper demodulator with the square-wave signal produced by the VCO, which also drives the ETF's heater. An off-chip capacitor $C_{int} = 1\mu F$ then integrates the output from the chopper demodulator. The loop locks when the average demodulator output becomes zero, producing a frequency corresponding to the f_{90} of the ETF.

At the f_{90} of ETF, the preamplifier phase shift, which spreads over process and temperature, has to be negligible compared to the ETF phase shift of $90°$. Therefore, compared to [19], where the readout amplifier had a bandwidth of 2.5 MHz, the preamplifier in [13] consisted of an open-loop four-stage amplifier with each

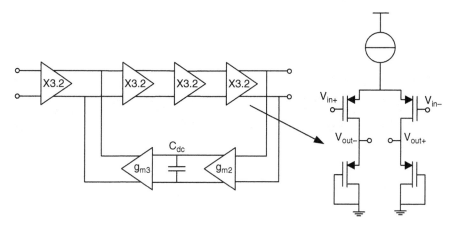

Fig. 3.24 Block-diagram of the thermopile preamplifier

stage made of a PMOS transconductor and PMOS diode loads. This configuration allowed for a bandwidth of more than 32 MHz. However, to prevent the output-referred offset of the preamplifier from saturating the synchronous demodulator, a DC servo loop consisting of a transconductance g_{m2}, a capacitor C_{dc} and a transconductor g_{m3} (Fig. 3.24), was built around it. This reduces the output-referred offset to the level of the offset of g_{m3}, but also gives the preamplifier a high-pass frequency response. To avoid introducing significant phase shift at around f_{90}, a high-pass corner frequency of a few Hz was chosen. This translates to a value of $C_{dc} = 100$ nF, requiring an off-chip capacitor.

As shown in Fig. 3.23 an extra pair of choppers is used after the synchronous demodulator [20]. These choppers are operated at a lower frequency of $f_{VCO}/128$, and their function is to chop the residual offset caused by the mismatched charge injection of the synchronous demodulator. As a consequence of adding these choppers, and in order to maintain the polarity of the loop, a low-frequency chopper should also be added at the input of the readout circuitry. This could have been placed at the output of the ETF, however, in order not to disturb the small signal at the thermopile output; this chopper was shifted through the ETF to the heater drive. The chopping of heat signal was done by periodically inversing the phase of the heater drive signal in phase with the low-frequency chopper after the demodulator.

The low-frequency chopping is a key contributor to the accuracy of this FLL, since any extra DC error added to the DC output of the synchronous demodulator, introduces an error to the phase detected by it, causing a frequency error. The low-frequency operation of the second pair of choppers prevents them from generating excessive residual offset. The low-frequency ripple produced by this action will then be filtered out by the large 1μF off-chip capacitor. Insufficient filtering of this ripple will cause excessive jitter in the VCO's output frequency.

The FLL of [13] achieved a device-to-device output frequency spread of ±0.25% (3σ) from −40°C to 105°C. The output frequency showed a $1/T^{1.7}$, with T the absolute temperature. This translated to a temperature sensing inaccuracy of

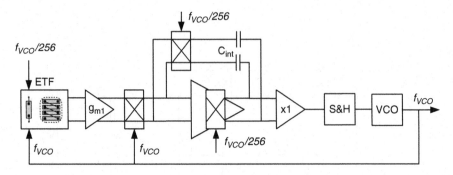

Fig. 3.25 Block-diagram of the electrothermal FLL with on-chip active integrator and off-chip sample-and-hold

$\pm 0.5°C$ (3σ). It can be seen from this FLL architecture that besides the large off-chip capacitor C_{dc} associated with the preamplifier, the large value of C_{int} is of extreme importance. On the one hand, it determines the 0.5 Hz noise bandwidth of the system, while on the other hand it filters out the low frequency chopping ripple caused by the chopped residual offset current of the synchronous demodulator.

A next iteration towards improving an electrothermal FLL was the implementation reported in [21]. In this work, the off-chip integrating capacitor was replaced by an on-chip active integrator around an op-amp (Fig. 3.25) and the effective transconductance of the readout was reduced by eliminating the preamplifier in [13]. This resulted in a noise bandwidth of 30 Hz with a transconductance of 100 μS and integrating capacitance of $C_{int} = 25$ pF. The elimination of the preamplifier allowed for a larger bandwidth; however, it increased the effect of the charge injection induced residual offset associated with the high frequency chopper demodulator.

The combination of low-frequency chopping and large residual offset in the FLL topologies of [13] and [21] leads to low frequency ripple. This ripple cause two major issues: firstly, it causes a large low-frequency term in the output frequency jitter, and secondly, and more importantly, it modulates the output frequency of the VCO. As a result, the low-frequency chopping clock will not have a 50% duty cycle. This causes residual offset, which significantly increases the inaccuracy of the output frequency.

To mitigate the problem of low-frequency ripple, an off-chip sample-and-hold was added before the VCO in [21]. However, its hold capacitors had values of 10 μF each and therefore, were implemented off-chip. Furthermore, addition of a sample-and-hold adds another low frequency pole to the loop due to the zero order hold function. If this pole is too close to the dominant pole of the loop, the loop can become unstable. The electrothermal FLL of [21] achieved an output frequency error of $\pm 0.45\%$ (3σ) from $-40°C$ to $100°C$ at an ETF power dissipation of 5 mW. This corresponds to a temperature sensing inaccuracy of $\pm 0.7°C$ (3σ).

3.8 Electrothermal FLL as Foundation for Frequency References

As discussed in the previous sections, electrothermal FLLs can be used to realize accurate temperature-to-frequency converters. Untrimmed device-to-device inaccuracies in the order of 0.25% (3σ) [13] show that the output frequency of such FLL is mainly determined by ETF characteristics. These are a function of its geometry and the thermal-diffusivity of silicon, which are both well-defined. Unlike the early thermal oscillators, where excess jitter and high power consumption of heater were major drawbacks, the narrow noise bandwidth provided by an FLL introduces significant improvement in both regards.

A frequency reference, in contrast to a temperature-to-frequency converter, requires the output frequency to be stable over temperature. Since the electrothermal FLL accurately locks the output frequency of the VCO to D, the output frequency will follow the same $T^{-1.8}$ characteristic of D (see Fig. 3.26). This corresponds to a temperature dependence of thousands of ppm/°C, which is not acceptable for a frequency reference. One solution to this is by measuring the temperature of the die, i.e. the temperature of the silicon substrate in which the ETF is implemented, and applying the temperature information to the FLL in order to temperature compensate it. This could be done by an absolute temperature sensor implemented next to the FLL (see Fig. 3.27).

The temperature information extracted by the absolute temperature sensor is mapped into a compensation signal and injected into the loop in order to keep the output frequency constant. It should be noted that the temperature sensing inaccuracy of the temperature sensor could then deteriorate the inherent accuracy of the FLL. However, state-of-the-art temperature sensors with inaccuracies in the range of 0.1°C have been reported [22]. This is still better than the 0.5°C temperature sensitivity of the electrothermal FLLs [13] and thus should provide the

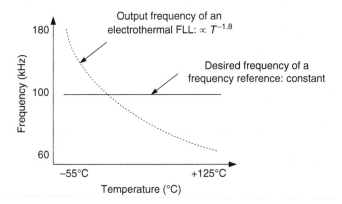

Fig. 3.26 Temperature dependent output frequency of an electrothermal FLL versus the desired constant output frequency of a reference

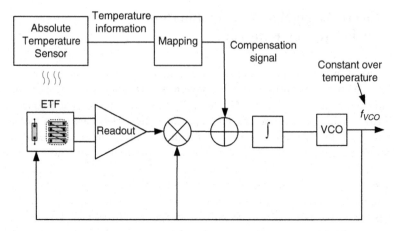

Fig. 3.27 Concept of temperature compensation applied to an electrothermal FLL for generation of a stable frequencies over temperature

possibility of accurately temperature compensating the loop. Building electrothermal frequency references based on electrothermal FLLs is the subject of the following chapters. Before progressing further on that topic, the dynamic behavior of FLLs will be discussed.

3.9 Dynamics of an Electrothermal FLL

In order to properly design and implement electrothermal FLLs, it is useful to analyze their dynamics and stability. Since an FLL locks the output frequency of a variable oscillator to the phase shift of an ETF, the main variables processed by the loop are phase and frequency, and so a brief review of these concepts will be first provided.

Figure 3.28a shows a sinusoidal signal $A(t) = A_m\sin(\omega_0 t)$ in the time domain, while Fig. 3.28b shows its phase, i.e. $\omega_0 t$, which is the total argument of the signal as a function of time. This is linearly changing with time with a slope of ω_0. The time-domain value of signal $A(t)$, therefore, crosses zero whenever its phase becomes an integer multiple of π. It can be seen that the larger the slope of the time domain value of the phase, the faster $A(t)$ zero-crossings will occur, i.e. the larger the frequency. Therefore, it can be concluded that the faster the phase of a signal changes, the higher its frequency will be. This means that the frequency term ω can be defined as the derivative of phase ϕ with respect to time [23]:

$$\omega = \frac{d\phi}{dt}. \tag{3.19}$$

Fig. 3.28 Phase of a sinusoid signal as a function of time

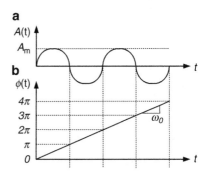

Fig. 3.29 The phase frequency relation of the input and output of an ETF

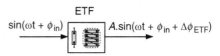

In order to simplify the study of ETF phase-frequency behavior, we will assume that the ETF consists of a point-heater source and a point-temperature sensor, where the heater is driven by a normalized sinusoidal signal. This could be justified by the fact that the square-wave heater drive signal will be low-passed filtered by the ETF. Figure 3.29 shows a simplified diagram of this ETF, where a time-domain input signal at frequency ω and initial phase ϕ_{in} is applied to the ETF. The output signal includes an excess phase shift $\Delta\phi_{ETF}$ caused by the thermal delay. Its amplitude, A, is a function of the thermal impedance of the structure [see (3.9)], the power dissipated in the heater, and the sensitivity of the relative temperature sensor. The excess phase shift in the output signal of an ETF equals:

$$\Delta\phi_{ETF} \propto s \cdot \sqrt{f \cdot T^{1.8}}. \tag{3.20}$$

where s is the distance between the point-heater and the point-temperature sensor, T is the absolute temperature, and f is $\omega/2\pi$. Therefore, the ETF phase shift can be written as a function of temperature and frequency:

$$\Delta\phi_{ETF} = R(T)\sqrt{f}. \tag{3.21}$$

where $R(T)$ is a temperature-dependent value determined by thermal diffusivity of silicon, D, and the ETF geometry.

With this simplified phase-domain model of an ETF, its input–output phase relationship can be written as:

$$\phi_{out} = \phi_{in} + \Delta\phi_{ETF}. \tag{3.22}$$

by inserting (3.21) into (3.22) we get:

$$\phi_{out} = \phi_{in} + R(T)\sqrt{f}. \tag{3.23}$$

Fig. 3.30 The simplified
block diagram of an
electrothermal FLL

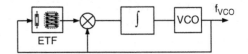

which considering (3.19) can be expanded to:

$$\phi_{out} = \phi_{in} + R(T)\sqrt{\frac{d\phi_{in}}{dt}}. \tag{3.24}$$

As shown by (3.24), to the first order, an ETF can be approximated as a two-port
block that receives a frequency at the input and provides an excess phase shift at the
output for a given temperature point.

As a further step towards simplification of the system, we assume its behavior to
be small signal around its steady state point on an ETF's phase-frequency characteristic (see Fig. 3.14). This assumption is correct to the first order because in a
locked electrothermal FLL, the loop steady-state normally changes only due to
thermal noise and the ambient temperature variations and these can be considered
to be small. This allows for further simplification of (3.24), through the use of
Taylor expansion of a square-root function, and assuming small values for $d\phi_{in}/dt$.
Hence, (3.24) can be simplified to:

$$\phi_{out} = \phi_{in} + R(T) \cdot \left(\frac{1}{2} + \frac{1}{2}\frac{d\phi_{in}}{dt}\right). \tag{3.25}$$

This shows that as far as small signal behavior is concerned, besides adding a
temperature-dependent phase shift, an ETF also behaves like a differentiator in the
phase domain.

To gain further insight into the system-level behavior of an FLL we consider the
simplified block-diagram of Fig. 3.30, where a synchronous demodulator plays the
role of phase detector in the loop. The single tone excitation model of an ETF
(Fig. 3.29) can be combined with a synchronous demodulator at its output, which
multiplies ETF output with the same signal that drives its heater (Fig. 3.31a). It can
be shown, through trigonometric analysis that the output of the synchronous
demodulator will be:

$$\begin{aligned}A \cdot \sin(\omega t + \phi_{in} + \Delta\phi_{ETF}) &\times \sin(\omega t + \phi_{in})\\ &= 0.5 \cdot A \cdot [\cos(\Delta\phi_{ETF}) - \cos(2\omega t + 2\phi_{in} + \Delta\phi_{ETF})]. \end{aligned} \tag{3.26}$$

From (3.26) it can be seen that this includes a DC term that is proportional to the
cosine function of $\Delta\phi_{ETF}$, and some high frequency content, which will be removed
by the loop filter (integrator) of the FLL. Therefore, the combination of an ETF and
a synchronous demodulator in the amplitude domain can be modeled in the phase
domain as an ETF and a phase detector whose output passes through a cosine
function.

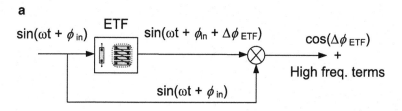

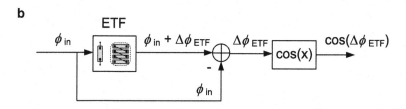

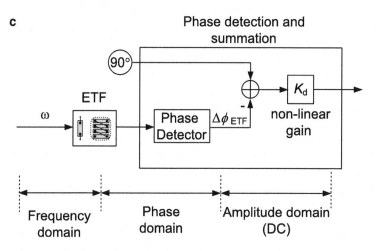

Fig. 3.31 The amplitude-domain simplified model of (**a**) an ETF with a synchronous demodulator, and (**b**) the phase-domain equivalent of that, and (**c**) a simplified phase summation node equivalent of the system

The cosine function can be seen as a non-linearity gain around 90° phase, introduced to the system (Fig. 3.31b). Therefore, we add this non-linear gain K_d to the loop [see (3.26)]. For a point-heater point-temperature sensor and at a given steady-state excitation frequency ω and temperature, K_d is determined by the heater power dissipation, P_{heater}, the ETF's temperature sensor sensitivity, k_{tp}, and the thermal impedance of the structure (defined by the distance s). This parameter should be determined numerically:

$$K_d \propto 0.5 \cdot \frac{1}{2\pi s} \cdot P_{heater} \cdot k_{tp} \cdot \exp\left(-s\sqrt{\frac{\omega}{2D}}\right) \cdot \cos\left(-s\sqrt{\frac{\omega}{2D}}\right). \qquad (3.27)$$

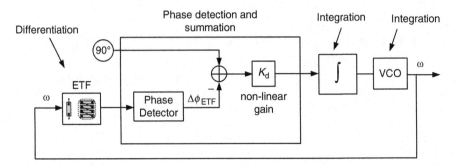

Fig. 3.32 The simplified phase domain approximated model representing a first-order system

With slight reconfiguration of the block-diagram in Fig. 3.31b it can be further simplified to that in Fig. 3.31c showing an ETF driven at a given frequency ω, and followed by a phase detector that can ideally extract its phase shift $\Delta\phi_{ETF}$. The cosine function shown in Fig. 3.31b can now be replaced by a summation node in phase-domain. This is followed by a gain K_d that produces an amplitude-domain signal (in DC) from the comparison of $\Delta\phi_{ETF}$ with a fixed phase reference of 90° in phase-domain.

Considering the abovementioned simplifications in the phase-domain, the block-diagram of an FLL (Fig. 3.30) can be approximated with a phase-domain feedback system including the model of Fig. 3.31c and including a summation node and a phase reference of 90° (Fig. 3.32). A VCO translates phase to frequency by means of integration. This means that the VCO is an integrator in the loop and therefore introduces a pole at DC. As shown by (3.25), considering the small signal behavior of the loop around its steady-state point, the ETF can be assumed as a block that receives frequency and produces phase at the output, which means a differentiator function. A differentiator introduces a zero at DC, and thus cancels the effect of the VCO's pole. This means that the order of the loop is determined by the loop filter, and is thus first order. It can be concluded that the dynamics of an electrothermal FLL approximately resembles that of a type I PLL, and therefore is very stable.

Besides providing insight to the dynamics of the loop, this simplified model of an electrothermal FLL allows for estimation of its noise-bandwidth. As shown further in this chapter, this is especially important for studying the jitter behavior of the loop. To further simplify the model, the combination of the ETF and phase detector shown in Fig. 3.32 can be replaced with a frequency-to-phase conversion gain K_ϕ [°/Hz]. This is like a sensitivity function and can be calculated numerically for the small-signal analysis of the loop around its steady-state. This can be done by considering a fixed temperature and a steady VCO frequency of ω_0 that correspond to an ETF thermal phase of 90°. The ETF then translates variations in its input frequency to variations in phase around this point. For the practical ETF structures, the sensitivity function, K_ϕ, should be determined via the thermal impedance model *numerically*. For the simple case of a point-heater point-temperature source structure, the steady state operation of the loop requires:

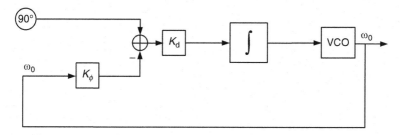

Fig. 3.33 Simplified phase-domain model of an electrothermal FLL that can be used for closed loop studies

$$\Delta\phi_{ETF} = -s\sqrt{\frac{\omega_0}{2D}} = 90°. \tag{3.28}$$

where D is the thermal diffusivity of silicon and s is the distance between the heater and the temperature sensor. For this model K_ϕ is simply calculated through the derivative of (3.28). The resulting final model, after addition of K_ϕ, is shown in Fig. 3.33.

3.10 FLL Behavioral Simulations

In order to confirm the validity of the simplified phase-domain model derived above, it is useful to investigate the dynamic behavior of the loop via its step response. To do this, a time-domain *Matlab* behavioral model of FLL shown in Fig. 3.30 can be made. The ETF can be modeled by the Foster network shown in Fig. 3.15 and with values provided in Table 3.2.

As shown in Fig. 3.34, the FLL is excited by a step pulse applied to the VCO input signal. The step will immediately cause the VCO output frequency to deviate from its steady-state value, but due to feedback the integrator will be driven such that f_{VCO} returns to the value corresponding to the 90° ETF phase shift.

To perform this simulation, an integrator with a unity gain frequency of ω_u = 300 kHz, a VCO with a sensitivity of K_{VCO} = 100 kHz/V, and the RC network representing the ETF characteristic are considered. The ETF model has f_{90} = 108 kHz, which will be the steady state locking frequency of the loop. Using 24 thermocouple arms for the thermopile with the sensitivity of each thermocouple being 0.5 mV/K, two different levels of heater power were simulated at 3 and 10 mW. This way the effect of the heater power on the predicted K_d parameter of the loop (see Fig. 3.32) can be verified. The simulated ETF output is shown in Fig. 3.35, which illustrates that the ETF outputs a larger signal for larger heater power dissipation (2.2 mV in contrast to less than 1 mV).

The time-domain simulation results of the FLL step response are summarized in Fig. 3.36. The startup of the loop from the initial zero condition can be seen at the

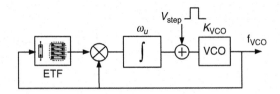

Fig. 3.34 Block diagram of the FLL model for step-response modeling

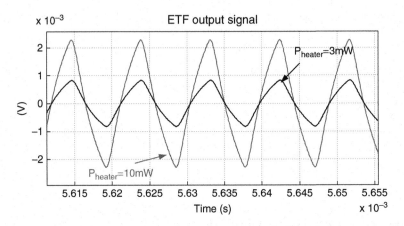

Fig. 3.35 Time-domain simulated ETF output signal at two heater power

beginning of the simulation (for proper scaling, the data near the start point has been omitted). It is clear that at higher heater power the FLL starts up faster. A step voltage with an absolute amplitude level of 200 mV has been applied to the input of VCO (see Fig. 3.36a). As a result of that, the frequency of the VCO changes instantaneously, which changes the ETF phase shift and hence the DC term at the output of the synchronous demodulator [see (3.26)]. This will be integrated by the integrator by controlling the VCO input voltage such that the frequency returns to the f_{90} of the ETF. The loop has to react again the same way when the step is removed.

The settling behavior of the loop as a response to the step (see the integrator output in Fig. 3.36c) shows pure *first order* behavior with a time constant which is a function of the heater power. The step response shows that the loop settles faster for the case of 10 mW heater power dissipation compared to 3 mW. This implies a larger parameter K_d, which means a larger loop-gain and thus wider bandwidth. From these step response simulations the closed-loop bandwidth can be calculated to be 240 and 80 Hz for the 10 and 3 mW simulations, respectively.

A further confirmation of the modeled behavior was sought in practice by applying the same experiment step signal to an electrothermal FLL and monitoring the integrator output settling to a new value with a first order response. The resulting measured step response from this experiment can be seen in Fig. 3.37,

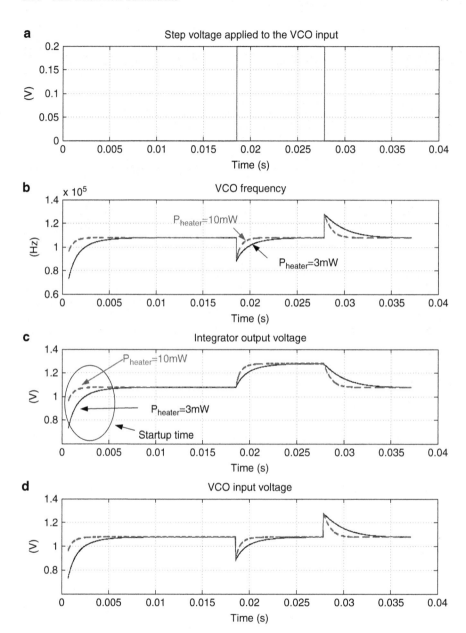

Fig. 3.36 Simulated FLL's step-response for two different heater powers: (**a**) the absolute value of the applied step to the loop, (**b**) the VCO's output frequency, (**c**) the integrator's output voltage, (**d**) the VCO's input voltage

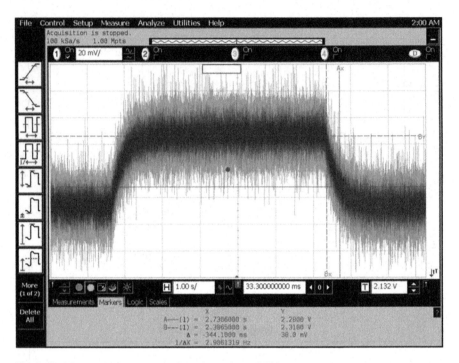

Fig. 3.37 Measured step response of an electrothermal FLL

which shows an approximate loop bandwidth of 1 Hz. This number is in line with those predicted by the simulations. The loop bandwidth in the simulations is more than 100 times larger because the unity-gain bandwidth of the integrator assumed for the simulations is 300 kHz, which is more than 100 times larger than that implemented in the practical loop. This choice of simulation conditions was made on purpose in order to limit the required simulation time and memory.

3.11 The Effect of Noise on an FLL's Jitter

3.11.1 ETF Noise

Section 3.2 mentioned that the major problem of the first generation thermal relaxation oscillators published in [6, 7] was the wide band thermal noise associated with the temperature sensor of their ETFs. This wide-band noise appears directly at the input of the oscillator's comparator and causes excessive jitter by introducing uncertainty at the moment the comparator decides. In later realizations [13], by incorporating an ETF in a feedback loop and using a synchronous demodulator as a narrow-band tracking filter, the effect of the ETF's thermal noise was significantly reduced.

Fig. 3.38 Inclusion of the ETF thermal noise in the electrothermal FLL

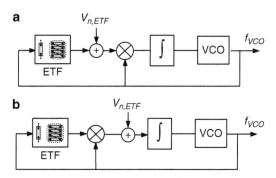

An ETF generates thermal noise due to the resistance of its p$^+$ diffusion thermocouple arms in the thermopile. For the ETF shown in Fig. 3.13, the total value of this resistance is in the order of 20 kΩ, which translates into a thermal noise density of about 18 nV/$\sqrt{Hz}$. This can be modeled as a noise source added to the output of the ETF, and therefore it can be included in the simplified block-diagram of an electrothermal FLL, as shown in Fig. 3.38a.

If the rest of the loop is ideally noise-free, the effect of the ETF's thermal noise, $V_{n,ETF}$, on the VCO output frequency can be further studied. Since to the first order, the multiplier is a chopper demodulator, its contribution to the noise transfer can be assumed negligible. Therefore, by applying the wide-band output noise of the ETF to (3.26) we can re-write the multiplier output as:

$$[A \cdot \sin(\omega t + \phi_{in} + \Delta\phi_{ETF}) + V_{n,ETF}] \times \sin(\omega t + \phi_{in}). \qquad (3.29)$$

With the assumption that an ETF's noise is mainly thermal noise (this is to first order valid since this noise is mainly generated by the resistance of the thermopile p$^+$ arms), the ETF noise can be shifted directly through the chopper demodulator (Fig. 3.38b), which means that it is added to the DC term at the output of the demodulator:

$$0.5 \cdot A \cdot \cos(\Delta\phi_{ETF}) + V_{n,ETF}. \qquad (3.30)$$

This noise is added directly to the FLL's error signal, which is the input to the integrator. The phase-domain model of an FLL shown in Fig. 3.33 can be adopted again for the noise-transfer analysis. This is done by adding a noise source $V_{n,ETF}$, originated by the ETF and shifted through the phase detector, at the input of the integrator (see Fig. 3.39).

We are interested in the transfer function relating the noise source $V_{n,ETF}$ to the output parameter ω of the loop. As can be seen from Fig. 3.39, this transfer needs to be calculated in a closed-loop system, where at the forward path the integrator is assumed to have a transfer function $H(s)$ and the VCO is assumed to have a voltage-to-frequency gain K_{VCO}. The feedback factor is determined by the ETF. The noise transfer of the FLL is defined by the small signal characteristics of the loop.

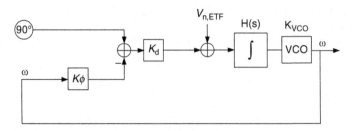

Fig. 3.39 Addition of ETF thermal noise to the simplified phase-domain model of the FLL

Therefore, the combined ETF and phase detector small signal transfer K_ϕ [°/Hz] should be determined numerically.

From the system block diagram of Fig. 3.39, the following noise transfer can be extracted:

$$T_{n,ETF}(s) = \frac{\omega_{n,ETF}(s)}{V_{n,ETF}(s)} = \frac{K_{VCO} \cdot H(s)}{1 + K_d \cdot K_\phi \cdot K_{VCO} \cdot H(s)}. \tag{3.31}$$

It should be noted that the loop parameters K_d and K_ϕ are dependent on the thermal impedance of the ETF, and thus need to be calculated numerically for a given steady-state point. From (3.31), the total noise at the VCO output frequency can be calculated as:

$$\omega_{no,ETF}^2 = \int_0^\infty \left| T_{n,ETF}(s) \right|^2 \cdot V_{n,ETF}^2 \cdot ds. \tag{3.32}$$

Since the integrator transfer function $H(s)$ is a low-pass function, and that the gains K_d and K_ϕ can be assumed to be approximately constant for fixed heater power, temperature, and s, the ETF thermal noise is low-pass filtered by the loop. This means that the total noise (jitter at the output frequency $\omega_{no,ETF}$) will be limited by the noise bandwidth of the system. This can be better seen from the bode-diagram showing the open as well as the closed-loop transfer $T_{n,ETF}(s)$. The loop bandwidth ω_{loop} is determined by the ETF in the feedback path and by the unity-gain bandwidth of the loop ω_u, which is determined by the integrator.

3.11.2 Implications for FLL Design

The total noise at the VCO output, i.e. the jitter of the output frequency, is determined by (3.32), and thus it is worth to consider the effect of FLL design parameters on the resultant output noise. This includes the ETF design as well as the loop design. In the first place, the total noise generated by the ETF thermopile can be reduced. This can be achieved by reducing its resistance. The other effect that an

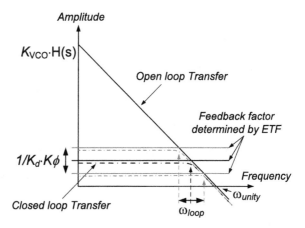

Fig. 3.40 Open and closed-loop transfer function of the simplified FLL model

ETF has on the loop jitter is its influence on the noise transfer, $T_{n,ETF}(s)$ [see (3.31)]. For large values of $H(s)$ (within the loop bandwidth), this is mainly determined by the effect of the ETF on the feedback factor of the loop via K_ϕ and K_d:

$$H(s) \to \infty \Rightarrow T_{n,ETF}(s) \approx \frac{1}{K_d \cdot K_\phi}. \tag{3.33}$$

Qualitatively, K_d is the more dominant parameter, since for small-signal behavior and in the steady-state condition, K_ϕ can be assumed to be constant. Therefore, the larger the ETF heater power dissipation, or the smaller the value of s [see (3.9) and (3.27)] the larger the value of K_d and thus, the smaller the noise transfer function. This means that as far as the effect of the ETF on the FLL jitter is concerned, either its heater power dissipation should be increased and/or its dimension (heater thermopile distance s) has to be decreased, and/or a thermopile should be adopted with a larger sensitivity (all leading to a larger signal at the ETF output).

Besides the effect of the ETF on the initial thermal noise level and the noise transfer, the other design parameter influencing the FLL jitter is the integrator unity gain frequency ω_u. Since the system has a first order transfer, the loop bandwidth determines its noise bandwidth [24]:

$$NBW = \frac{\pi}{2}\omega_{loop}. \tag{3.34}$$

where ω_{loop} can be determined considering Fig. 3.40, (3.31) and (3.33):

$$K_{VCO} \cdot \omega_u = \omega_{loop} \cdot \frac{1}{K_d \cdot K_\phi} \Rightarrow \omega_{loop} = K_d \cdot K_\phi \cdot K_{VCO} \cdot \omega_u. \tag{3.35}$$

Therefore (3.32) can be simplified to:

$$\omega^2_{no,ETF} = V^2_{n,ETF} \cdot \left| \frac{1}{K_d K_\phi} \right|^2 \cdot NBW$$

$$= V^2_{n,ETF} \cdot \left| \frac{1}{K_d K_\phi} \right|^2 \cdot \frac{\pi}{2} \cdot K_d \cdot K_\phi \cdot K_{VCO} \cdot \omega_u \qquad (3.36)$$

$$= \frac{\pi}{2} \cdot V^2_{n,ETF} \cdot \frac{K_{VCO} \cdot \omega_u}{K_d K_\phi}.$$

Example consider a point-heater point-temperature sensor ETF structure with s = 10 μm, a heater power dissipation P_{diss} = 10 mW, a thermocouple temperature sensor with a sensitivity of 0.5 mV/K, and a thermal noise of 20 nV/√Hz at a loop steady-state frequency of ω_0 = 285 kHz (see Fig. 3.3) associated with a phase shift of 90° for this structure. The loop incorporates a VCO with a sensitivity of K_{VCO} = 250 kHz/V, producing a single-tone sinusoid signal and an integrator with a unity gain frequency of 500 Hz.

The parameter K_ϕ of the loop can be calculated numerically around the steady-state point of the loop, by considering the phase characteristic of the thermal impedance of the structure (Fig. 3.3). This can be done once the thermal impedance of the structure has been calculated using (3.9) and numerically taking the derivative of this response. For the specifications mentioned, the calculated K_ϕ is approximately 0.16 [°/kHz].

Furthermore, the parameter K_d can be calculated from (3.9) and (3.26). The thermal impedance of the structure (3.9) predicts a temperature variation of about 0.4 K as a function of a 10 mW power dissipation at the heater. For the thermocouple sensitivity mentioned, this leads to an output signal amplitude of about 0.2 mV. With (3.26) and the DC value resulting from the phase detector, the value of K_d will be 0.1 [mV/°]. This leads to a $K_d.K_\phi$ in the order of 0.016 [mV/kHz]. The loop noise bandwidth therefore can be calculated from (3.34) and (3.35) to be about 3Hz. Through (3.36) for a $V_{n,ETF}$ = 20 nV/√Hz the expected rms jitter at the output frequency of FLL will be <100 ps. Compared to the steady-state locking frequency of ω_0 = 285 kHz, this jitter accounts for 0.003% of the oscillation period, which can be acceptable.

3.11.3 VCO Noise

The other noise source in the system is the inherent noise of the VCO. The VCO is a block that translates a control voltage into an output frequency with a certain gain, K_{VCO}. Like any other electronic circuit, a VCO can be assumed to be a noise-less element and its total circuit noise can be derived from its input as a noise source $V_{n,VCO}$ (see Fig. 3.41). This results in a different noise transfer for the VCO noise contribution to the FLL output frequency:

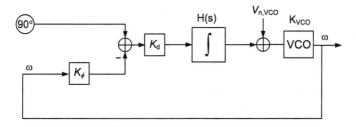

Fig. 3.41 The introduction of the VCO noise into the simplified model of an FLL

$$T_{n,VCO}(s) = \frac{\omega_{n,VCO}(s)}{V_{n,VCO}(s)} = \frac{K_{VCO}}{1 + K_d \cdot K_\phi \cdot K_{VCO} \cdot H(s)}. \tag{3.37}$$

This transfer function suggests that for a low-pass characteristic associated with $H(s)$, the VCO noise is high-pass filtered. This means that to minimize the VCO noise contribution to the FLL jitter, the loop bandwidth should be as large as possible. This can be seen intuitively from (3.37), where the larger $H(s)$, the smaller the ω_n due to the $V_{n,VCO}$. Therefore, the larger the loop gain at higher frequencies, the more the suppression of VCO noise. This is however, in contradiction with the previous conclusion regarding the ETF thermal noise effect on the FLL's output jitter. That analysis required a smaller loop bandwidth for less output noise.

Considering that the ETF noise and the VCO noise are the only major noise sources in the system, we can write:

$$\omega_{no} = \left[\omega_{no,ETF}^2 + \omega_{no,VCO}^2\right]^{0.5}. \tag{3.38}$$

Therefore, when the ETF is the dominant source of noise, the loop bandwidth should be minimized, however if an ETF is designed with a low output thermal noise, to reduce the VCO's noise effect, the loop bandwidth can be increased.

It should also be noted that our definition of noise at the output frequency of the VCO is analogous to the cycle-to-cycle jitter's definition. This is simply defined as the fluctuations in a VCO's output signal period of oscillation referred to a fixed single edge of the signal as a reference.

3.12   Challenges Associated with the Previous FLL's

From the previous review of the work done on electrothermal FLLs it can be concluded that the key to achieving reasonable jitter performance at acceptable levels of heater power dissipation is the use of a synchronous demodulator in combination with an integrator in a feedback loop. In contrast to the earlier thermal oscillators, this approach provides a narrowband solution that limits the effect of ETF thermal noise on the frequency jitter. Therefore, when making an on-chip frequency reference based on a thermal delay, an electrothermal FLL is the most promising choice.

Before we progress further, it is important to mention the major challenges associated with the FLL architectures introduced due to their use of analog signal processing. The major challenge when implementing such FLLs is the required narrow noise bandwidth of the loop, which is associated with the required jitter performance. This translates into large time constants for the loop integrator, which in turn translates into large capacitance values for the circuitry implementing this block. Therefore the prior architectures are less amenable to CMOS integration, such as the large 1 µF off-chip capacitor in [13] and the 10 µF off-chip sample-and-hold capacitors in [21]. This calls for the need to adopt a different topology for an electrothermal FLL, which is the topic of the next chapter.

3.13   Conclusions

Integrated frequency references have utilized the electrical, mechanical and magnetic properties of on-chip components. However, the thermal properties of silicon have hardly been explored. Heat generated in a silicon substrate causes temperature variations. The generated heat diffuses at a defined rate through the substrate, which means that the resulting temperature fluctuations can be sensed at a further distance after a defined delay. This is the basis of implementing on-chip thermal delays. These are defined by the thermal diffusivity of silicon and the distance between the heat source and heat sensing points on the substrate. Due to the accuracy of lithography and the purity of IC grade silicon, these delays are well defined.

The thermal diffusivity of silicon can be harnessed by a CMOS compatible element called an electrothermal filter (ETF). An ETF made of a heater and a relative temperature sensor, located at a distance s apart, behaves like a low pass filter. This filter has a well-defined phase response, which is determined by the thermal diffusivity of silicon and the ETF geometry. The major challenge in interfacing ETFs is their small output signals.

Thermal delays were used in early thermal feedback oscillators, which measured the silicon die temperature with a temperature-dependent frequency. The temperature dependence of the thermal delay is due to that of the thermal diffusivity D of silicon. These oscillators had a rather poor jitter performance, accounting for tens of percent of the output period even at large levels of heater power dissipation. This was because of the thermal noise associated with their thermal delay line.

The jitter problem of the early thermal oscillators was solved by embedding an ETF in an electrothermal frequency-locked loop (FLL). In this type of loop, feedback locks the output frequency of a VCO to the phase shift of an ETF. The loop incorporates a synchronous demodulator and an integrator that together form a narrow-band tracking filter. This helps filter the wide-band thermal noise associated with the relative temperature sensor of the ETF, allowing reasonable jitter to be achieved at relatively low heater power dissipations. This makes an electrothermal

FLL the architecture of choice for building a thermal diffusivity-based frequency reference.

The ultimate jitter performance of an FLL is determined by the ETF level of thermal noise, as well as the loop noise bandwidth. In order to achieve narrow noise bandwidths, early FLLs utilized large time constants. This required large capacitors that make the CMOS integration of an FLL difficult. This calls for a new architecture to solve this problem.

References

1. Hamann HF et al (2007) Hotspot-limited microprocessors: direct temperature and power distribution measurements. IEEE J Solid-State Circ 42(1):56–65
2. Trimmer W (1997) Micromechanics and MEMs: classic and seminal papers to 1990. IEEE publications, New York, pp 353–433. ISBN 9780470545263
3. Gray PR, Hamilton DJ (1971) Analysis of electrothermal integrated circuits. IEEE J Solid-State Circ 6(1):8–14
4. Matzen WT et al (1964) Thermal techniques as applied to functional electronic blocks. Proc IEEE 2:1496–1501
5. Friedman MF (1969) Monolithic high-Q bandpass filters employing electrothermal circuits. Ph.D. dissertation, Department of Electrical and Computer Engineering, University of Arizona, Tucson
6. Szekely V (1994) Thermal monitoring of microelectronic structures. Microelectron J 25(3):157–170
7. Szekely V et al (1995) A new monolithic temperature sensor: the thermal feedback oscillator. In: Proceedings of the transducers, Stockholm, Sweden, June 1995, pp 124–127
8. Bosch G (1972) A thermal oscillator using the thermo-electric (seebeck) effect in silicon. Solid State Electron 15(8):849–852
9. Turkes P (1983) An ion-implanted resistor as thermal transient sensor for the determination of the thermal diffusivity in silicon. Physica Status Solidi A 75(2):519–523
10. Touloukian YS et al (1998) Thermophysical properties of matter, vol 10. Plenum, New York
11. Vermeersch B (2009) Thermal AC modelling, simulation and experimental analysis of microelectronic structures including nanoscale and high-speed effects. Ph.D. dissertation, University of Gent
12. Veijola T (1996)"Simple model for thermal spreading impedance. In: Proceedings of the BEC'96, Tallinn, Estonia, October 1996, pp 73–76
13. Makinwa KAA, Snoeij MF (2006) A CMOS temperature-to-frequency converter with an inaccuracy of less than ±0.5 °C (3σ) from −40 °C to 105 °C. IEEE J Solid-State Circ 41(12):2992–2997
14. van Herwaarden AW, Sarro PM (1986) Thermal sensors based on the seebeck effect. Sens Actuat 10:321–346
15. Xia S, Makinwa KAA (2007) Design of an optimized electrothermal filter for a temperature-to-frequency converter. In: Proceedings of the IEEE sensors, Atlanta, GA, October 2007, pp 1255–1258
16. Kashmiri SM et al (2009) A temperature-to-digital converter based on an optimized electrothermal filter. IEEE J Solid-State Circ 44(7):2026–2035
17. Makinwa KAA (2004) Flow sensing with thermal sigma-delta modulators. Ph.D. dissertation, Delft University of Technology, Delft
18. Hirai T, Asai T, Amemiya Y (2010) A CMOS phase-shift oscillator based on the conduction of heat. J Circuit Syst Comput 19(4):763–772

19. Makinwa KAA, Witte JF (2005) A temperature sensor based on a thermal oscillator. In: Proceedings of the IEEE sensors, Irvine, CA, pp 1149–1152
20. Bakker A, Huijsing JH (1996) Micropower CMOS temperature sensor with digital output. IEEE J Solid-State Circ 31(7):933–937
21. Zhang C, Makinwa KAA (2008) Interface electronics for a CMOS electrothermal frequency-locked-loop. IEEE J Solid-State Circ 43(7):1603–1608
22. Pertijs MAP et al (2005) A CMOS smart temperature sensor with a 3σ inaccuracy of ± 0.1 °C from -55 °C to 125 °C. IEEE J Solid-State Circ 40(12):2805–2815
23. Razavi B (2001) Design of analog CMOS integrated circuits. McGraw-Hill, New York
24. Johns DA, Martin K (1997) Analog integrated circuit design. Wiley, New York

Chapter 4
A Digitally-Assisted Electrothermal Frequency-Locked Loop in Standard CMOS

This chapter describes an electrothermal frequency-locked loop (FLL) that is suitable for CMOS integration. An electrothermal FLL requires a narrow noise-bandwidth to limit the jitter resulting from the thermal noise of its electrothermal filter (ETF). This is rather challenging to implement in the analog domain, since the narrow bandwidth requires the realization of a large time constant. This chapter proposes a digitally-assisted FLL (DAFLL) architecture that mitigates the integration difficulties of previous FLLs. In the DAFLL, the narrow-band loop filter is realized in the digital domain. As such, it does not require off-chip analog components. Initially, the proposed system-level architecture will be introduced. Later, the design, implementation and characterization of the major building blocks will be covered. These include a phase digitizer realized by means of a phase-domain $\Delta\Sigma$ modulator (PD$\Delta\Sigma$M), and a digitally-controlled oscillator (DCO).

The DAFLL is realized in two steps, each one of them resulting in a test chip. The first chip consists of two ETFs and a PD$\Delta\Sigma$M, which demonstrates that the phase shift of an ETF can be accurately digitized. It also enables a comparison between the two ETFs. The second chip builds on the results from the first, and realizes the complete DAFLL. Besides forming the basis of an electrothermal frequency reference, a DAFLL can also be used as a test-vehicle for characterizing the phase-frequency characteristic of an ETF. The final section of this chapter shows how this concept can be used in measuring the thermal-diffusivity of a chip.

4.1 Introduction

The previous chapter concluded that to implement a frequency reference, a temperature-compensated electrothermal FLL is the most suitable architecture. However, previous implementations required the realization of large time constants, which either require a very small transconductance and/or a very large

S.M. Kashmiri and K.A.A. Makinwa, *Electrothermal Frequency*
References in Standard CMOS, Analog Circuits and Signal Processing,
DOI 10.1007/978-1-4614-6473-0_4, © Springer Science+Business Media New York 2013

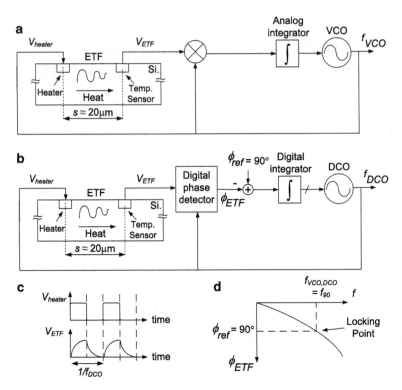

Fig. 4.1 (a) The analog FLL, (b) the digitally-assisted FLL(DAFLL), (c) the heater drive V_{heater} and the output signal V_{ETF} (d) the FLL's locking point

capacitor (see Chap. 3). The former solution leads to excess noise and a small signal both leading to extra jitter. The latter solution is not efficient due to the large amount of chip area.

A solution to the problem described above is to use a digital integrator. The implementation of the required time constant then translates into a choice of the digital integrator's sampling rate and the total number of bits it accumulates. The digital hardware required to implement this, occupies much less area than the realization of large capacitors [1, 2].

The simplified block diagrams shown in Fig. 4.1 compare an analog FLL [3] with the proposed digitally-assisted FLL (DAFLL) [4]. Figure 4.1a shows the analog loop including a VCO, an ETF, a synchronous demodulator and an analog integrator. The ETF's heater is driven by the VCO's output signal, V_{heater}, which results in an AC output signal V_{ETF} (Fig 4.1c). The analog FLL is locked when the demodulator's DC output is zero, corresponding to a VCO frequency $f_{VCO} = f_{90}$. At this frequency, the phase shift of the ETF is about 90°.

Figure 4.1b shows the digitally-assisted loop. Here, the ETF's phase shift, ϕ_{ETF}, is digitized by a digital phase detector. This is compared to a digital phase reference, ϕ_{ref}, at a phase summation node. The result of this comparison is an error signal, which is integrated by a digital integrator. The integrator drives a digitally-controlled oscillator (DCO) that, in turn, drives the ETF at frequency f_{DCO}. Once more, feedback forces the DCO to oscillate at a frequency, f_{DCO}, where $\phi_{ETF} = \phi_{ref}$ (Fig. 4.1d).

4.2 Proposing a Digitally-Assisted FLL

4.2.1 Operating Principle

A system-level block diagram of the proposed digitally-assisted FLL (DAFLL) is shown in Fig. 4.2. Due to the utilization of a digital filter, analog-to-digital and digital-to-analog conversions need to be considered in the loop. Here, the analog signal is the ETF phase, ϕ_{ETF}, which needs to be digitized. The digital output of the digital filter should then be translated to an output frequency by the DCO.

For a given ETF structure, ϕ_{ETF} is a function of the temperature and the drive frequency. Furthermore, ϕ_{ETF} is contained in the sub-millivolt output signal of the thermopile (see Chap. 3), and this needs to be digitized in the presence of wide-band thermal noise due to the thermopile resistance. Since for a given driving frequency, this phase shift varies only as a function of temperature, its variation has a bandwidth in the order of a few tenths of a Hz. Digitizing such a low-frequency signal with high resolution (typically >12-bits) matches the characteristics of delta-sigma ($\Delta\Sigma$) ADCs [5, 6]. Such ADCs exchange resolution in amplitude with resolution in time, by means of oversampling and noise shaping [5].

To digitize ϕ_{ETF}, a phase-domain delta sigma ($\Delta\Sigma$) modulator (PD$\Delta\Sigma$M) has been adopted to be embedded in the DAFLL. A PD$\Delta\Sigma$M is similar to a conventional amplitude-domain $\Delta\Sigma$ modulator [5], with the difference that the input and the reference values are not in the amplitude domain, but in the phase domain. The PD$\Delta\Sigma$M produces a bitstream output which represents ϕ_{ETF} in digital domain. Within the DAFLL, this bitstream is then digitally compared with a phase reference of 90° (see Fig. 4.2), and the resulting error signal is integrated by the digital filter.

The multi-bit output of the digital filter drives a digital-to-analog converter (DAC), whose output then drives a VCO. The combination of the DAC and the VCO forms a DCO. The PD$\Delta\Sigma$M, the digital filter, and the DAC are all

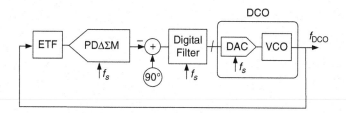

Fig. 4.2 Simplified system-level block-diagram of the DAFLL

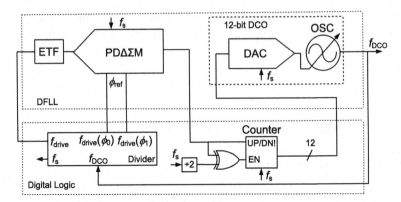

Fig. 4.3 Detailed block-diagram of the digitally-assisted electrothermal FLL

sampled at the same frequency f_s, which is a sub-multiple of f_{DCO}. The result is a self-referenced system, whose noise-bandwidth is determined by the digital filter, which can be almost arbitrarily low.

4.2.2 DAFLL System-Level Specifications

In order to derive a set of system-level specifications for the proposed DAFLL, the more detailed block-diagram shown in Fig. 4.3 is considered. The square-wave output of the DCO, f_{DCO}, enters a frequency divider. This produces the sampling clock, f_s, and the ETF drive signal, f_{drive}, which are both sub-multiples of f_{DCO}. The ETF output is then applied to the PD$\Delta\Sigma$M, which digitizes the ETF phase shift with reference to two phase references, $f_{drive}(\phi_0)$ or $f_{drive}(\phi_1)$, produced by the divider. These are also square-wave signals, which are $\pm 45°$ phase-shifted regarding to f_{drive} [7, 8]. The bitstream output of the modulator is then a digital representation of the ETF's phase shift.

In order to implement $\pm 45°$ phase references with a 50% duty cycle, $f_{DCO} = 16 \cdot f_{drive}$. The 90° phase reference of the FLL is a square-wave with a frequency of $f_s/2$ (corresponding to consecutive ones and zeros at rate f_s). This phase reference is subtracted from the PD$\Delta\Sigma$M bitstream output, and the resulting 3-level signal is integrated by an up/down counter. The counter acts like a digital integrator, and thus forms the loop's digital filter. When the 3-level signal is zero, the XOR gate disables the counter; otherwise the counter's state is appropriately incremented or decremented. The counter's output is then fed to the DCO.

Since the inaccuracy of a previous FLL [3] was 0.25%, the same level of inaccuracy is expected from the DAFLL. This sets the error budget of the loop's building blocks. A potential error source is the electrical domain phase error introduced by the PD$\Delta\Sigma$M and the ETF. The ETF is designed to have an f_{90} of about 100 kHz at room temperature and a lithography-limited thermal phase spread of 0.12° (3σ) in a 0.7 μm process [7]. Therefore, the PD$\Delta\Sigma$M's phase error should be designed to be much (10 times) lower than this. For this specification, a phase error of 0.01° will lead to an output frequency error of less than 0.04%, which is more than 5× smaller than the expected accuracy.

It was shown in Chap. 3 that due to the temperature dependence of D, the ETF's phase shift will be a function of temperature. Therefore, the DCO needs to have enough tuning range over the temperature. It should be noted that the DAFLL proposed in this chapter will not be temperature-compensated. This is mainly to confirm from the experimental results that the output frequency of the loop is indeed locked to the thermal diffusivity of silicon, D, i.e. follows the same $T^{-1.8}$ temperature dependence as was shown by the analog FLLs. This sets the specifications regarding the DCO range.

With the room temperature f_{90} of about 100 kHz and to produce the required phase references of the PD$\Delta\Sigma$M, $f_{DCO} = 16 \cdot f_{drive} = 1.6$ MHz (at room temperature). This means that the variations of the FLL's frequency over the military range (with $\phi_{ref} = 90°$) will require a DCO tuning range from 800 kHz to 3.2 MHz. In line with the expected accuracy of 0.25%, a frequency resolution of 0.05% was chosen. With the required DCO range over temperature and the extra ±40% additional spread, due to process and temperature variations, a 12-bit DCO with a step size of 800 Hz was devised.

The noise bandwidth of the system is determined by the length of the counter and the value of f_s. For a 12-bit counter and $f_s = f_{drive}/64$, the DAFLL's expected noise bandwidth is 0.4 Hz (at room temperature and $f_{drive} = 100$ kHz), which is close to the previous FLL reported in [3]. This ensures sufficient suppression of the wideband noise of the ETF and the quantization noise of the PD$\Delta\Sigma$M.

The expected resolution of the PD$\Delta\Sigma$M, within the full-scale phase range of ETF over temperature, should be chosen in line with the expected 0.25% inaccuracy level of the DAFLL. Therefore, for a frequency resolution of 0.05% a phase resolution of better than 20 milli-degrees should be sufficient (see (3.11)) showing the phase-frequency sensitivity of an ETF). For an ETF full-scale phase range of about 50° over the military temperature range, and at a constant drive frequency of about 100 kHz, this translates to a phase resolution of about 11.5 bits. Considering that within the narrow noise-bandwidth of the DAFLL, the oversampling ratio of the PD$\Delta\Sigma$M can be made quite large, a first-order $\Delta\Sigma$ modulator should be sufficient to achieve a thermal noise limited resolution better than 12 bits.

4.2.3   DAFLL Realization Phases

As mentioned earlier, in order to minimize risks during the development, the design of the DAFLL was split into two steps, each involving a separate test chip. The first chip only includes two ETFs and the PD$\Delta\Sigma$M, while the second chip realizes the complete DAFLL. The circuit and system design, as well as the experimental results of these two chips will be provided in two separate sub-sections in this chapter.

The first chip was to investigate the accuracy with which the phase shift of an ETF could be digitized. Also, a choice had to be made between the bar ETF used in [3] (shown in Fig. 3.11) and the optimized ETF (shown in Fig. 3.13). As described in Chap. 3, a greater accuracy was expected from the optimized ETF, which needed to be verified by experimental results.

The first chip, besides providing a development ground for the DAFLL, can also be used as a temperature-to-digital converter (TDC) [8, 9]. This is because the digitized ϕ_{ETF} is a function of temperature when the ETF is driven at a constant frequency, i.e. by a crystal oscillator. This was considered during the system and circuit design of this chip, which will be described in the following sub-sections. Furthermore, to add more experimental data to available performance of the thermal-diffusivity-based temperature sensors at the time [8], performance of the first chip was also evaluated as a TDC.

The second chip, builds on the results obtained from the first chip. The selected ETF in combination with the PD$\Delta\Sigma$M were integrated with a 12-bit DCO, which was designed for the required range and resolution described in the previous sub-section. The system and circuit design of the second chip, forming the complete DAFLL, as well as experimental results on characteristic and accuracy of its output frequency over temperature will be provided in a separate sub-section.

4.3   First Test Chip

The following sub-section will provide discussions on the system and circuit design of the first test chip. At the end, the experimental results from this chip and the conclusions based on those will be provided.

4.3.1 PDΔΣM System-Level Architecture

The first test chip implements ETFs and a PD$\Delta\Sigma$M [7]. This chip can also be used as a temperature-to-digital converter (TDC) [8, 9]. This is because the digitized ϕ_{ETF} is a function of temperature when the ETF is driven at a constant frequency, i.e. by a crystal oscillator. For this reason, the following paragraphs also consider the system-level aspects of the combination of an ETF and a PD$\Delta\Sigma$M as a TDC. This means that the error contribution of the various building blocks in the first chip's system will also be translated into temperature sensing inaccuracies. The main reason to consider temperature sensing is to facilitate a comparison between the performance of the optimized ETF (see Chap. 3) and the performance of the bar ETF, since the latter had previously been used in a similar TDC [8].

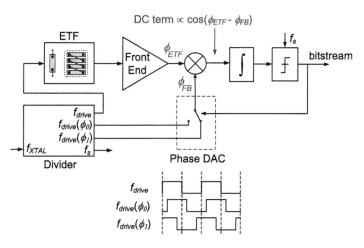

Fig. 4.4 Simplified block-diagram of the first test chip

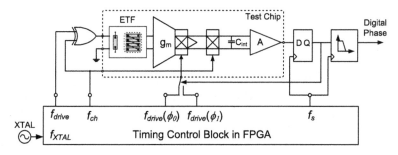

Fig. 4.5 Complete block diagram of the first test chip

A simplified block-diagram of the first test chip is shown in Fig. 4.4. The ETF is driven by a square-wave, at a constant frequency f_{drive}, which is derived by a frequency divider from a crystal oscillator oscillating at f_{XTAL}. The ETF's output, with a temperature-dependent phase-shift ϕ_{ETF}, is applied to a front-end. The output of the front-end is fed to a PD$\Delta\Sigma$M, which consists of a multiplier, an integrator, a quantizer and a single-bit phase DAC. Depending on the output of the quantizer, the output of the front-end will be multiplied by one of the two digitally phase-shifted versions of f_{drive}, $f_{drive}(\phi_0)$ and $f_{drive}(\phi_1)$, generated by the phase DAC. The multiplier acts as the $\Delta\Sigma$ modulator's summing node [5], and outputs a current whose DC component is proportional to the cosine of the phase difference between the output of the front-end, ϕ_{ETF}, and that of the phase DAC, ϕ_{FB}. The average of the modulator's bit-stream is a weighted average of the two reference phase shifts ϕ_0 and ϕ_1 that approximate ϕ_{ETF}.

A complete block diagram of the first test chip is shown in Fig. 4.5. The ETF is driven at $f_{drive} = 85$ kHz. Its sub-millivolt AC output signal is converted into a current by a transconductor g_m. The modulator's multiplier (Fig. 4.4) is implemented

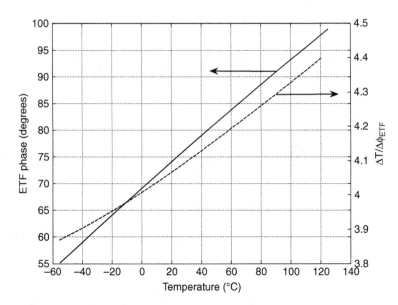

Fig. 4.6 Simulated phase shift of the ETF (*left axis*) and the resulting sensitivity-function relating temperature variations to phase variations (*right axis*)

as a chopper demodulator, embedded in the transconductor and its integrator is implemented by capacitor C_{int}. This capacitor also filters out the harmonics of f_{drive} present at the output of the demodulator. The voltage across C_{int} is then boosted by a differential-to-single-ended amplifier, and applied to a digital latch (in an FPGA). The latch acts as the modulator's quantizer. The FPGA also generates the two phase reference signals, $f_{drive}(\phi_0)$ and $f_{drive}(\phi_1)$ from the crystal-oscillator output.

For $f_{drive} = 85$ kHz, Fig. 4.6 illustrates the simulated phase shift of the optimized ETF over the military temperature range (left axis). This is a simulation based on the thermal impedance model that was introduced in Chap. 3. The derivative of this function's inverse is also shown. When the chip is used as a temperature sensor, the maximum value of this derivative defines a worst-case sensitivity factor $S^T_{\phi_{ETF}}$ that links ETF phase errors to temperature sensing errors:

$$S^T_{\phi_{ETF}} = \frac{\partial T}{\partial \phi_{ETF}} = 4.5 \left[\frac{°C}{degrees} \right]. \tag{4.1}$$

$$\Delta T = 4.5 \Delta \phi_{ETF}. \tag{4.2}$$

Therefore, ETF phase spread $\Delta \phi_{ETF}$, caused by lithographic inaccuracy (see Chap. 3), will give rise to temperature measurement errors. Errors contributed by other sources, such as the electrical phase spread added by the transconductor g_m and the residual offset added by the synchronous demodulator, should be made much smaller than this.

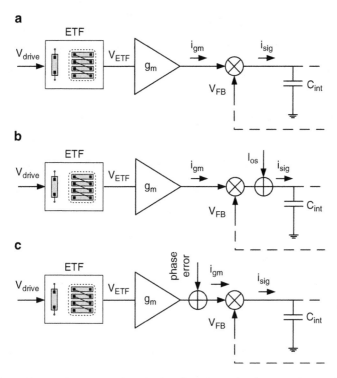

Fig. 4.7 Simplified block-diagram of (**a**) the interface electronics for an ETF (**b**), with residual offset, and (**c**) with phase error

An idealized block diagram of the first test chip front-end, based on the single tone excitation model introduced in Chap. 3, is shown in Fig. 4.7. This diagram shows the effect of error sources such as residual offset and excess electrical phase error. The harmonics present at the output of the synchronous demodulator are filtered out by the integrator of the phase-domain $\Delta\Sigma$ modulator. To first-order, therefore, the operation of the front-end and the effect of error sources on it can be analyzed by modeling both the ETF drive signal, V_{drive}, and the feedback signal of the phase-domain $\Delta\Sigma$ modulator, V_{FB}, as sinusoidal functions:

$$V_{drive}(t) = \cos(2\pi f_{drive} t). \tag{4.3}$$

$$V_{FB}(t) = \cos(2\pi f_{drive} t + \phi_{FB}). \tag{4.4}$$

where ϕ_{FB}, toggles between ϕ_0 and ϕ_1 depending on the polarity of the modulator's bit-stream. The ETF's output signal, V_{ETF}, will then be a phase-shifted version of V_{drive}:

$$V_{ETF}(t) = A \cdot \cos(2\pi f_{drive} t + \phi_{ETF}). \tag{4.5}$$

where A is the amplitude and ϕ_{ETF} is ETF's temperature-dependent phase shift. The amplitude A is a function of the ETF's geometry, the power dissipated in the heater, and the sensitivity of the thermopile. The transconductor g_m converts V_{ETF} into a current i_{gm}, which is then multiplied by the feedback signal, V_{FB}, leading to a demodulated signal i_{sig}:

$$i_{gm}(t) = g_m \cdot A \cdot \cos(2\pi f_{drive}t + \phi_{ETF}). \tag{4.6}$$

$$\begin{aligned} i_{sig}(t) &= i_{gm} \cdot V_{FB} \\ &= g_m \cdot A \cdot \cos(2\pi f_{drive}t + \phi_{ETF}) \cdot \cos(2\pi f_{drive}t + \phi_{FB}) \end{aligned} \tag{4.7}$$

The current defined by (4.7) has AC terms that will be filtered out by C_{int}, as well as a DC term that carries the phase information:

$$i_{sig,DC}(t) = \frac{1}{2} \cdot g_m \cdot A \cdot \cos(\phi_{ETF} - \phi_{FB}). \tag{4.8}$$

The feedback in the phase-domain $\Sigma\Delta$ modulator (Fig. 4.4) acts to balance the average charge accumulated by the integrator:

$$\begin{aligned} &\mu \cdot \frac{1}{2} \cdot g_m \cdot A \cdot \cos(\phi_{ETF} - \phi_0) \\ &+ (1 - \mu) \cdot \frac{1}{2} \cdot g_m \cdot A \cdot \cos(\phi_{ETF} - \phi_1) \approx 0. \end{aligned} \tag{4.9}$$

where μ is a number between 0 and 1, which represents ϕ_{ETF} as a weighted average of the two phase references, ϕ_0, and ϕ_1:

$$\mu \approx \frac{\cos(\phi_{ETF} - \phi_1)}{\cos(\phi_{ETF} - \phi_1) - \cos(\phi_{ETF} - \phi_0)}. \tag{4.10}$$

If $(\phi_{ETF} - \phi_1)$ and $(\phi_{ETF} - \phi_0)$ are close to 90°, the cosine function can be linearized, leading to a linear relation between μ and ϕ_{ETF}:

$$\mu \approx \frac{\phi_{ETF} - (90 + \phi_1)}{\phi_0 - \phi_1}. \tag{4.11}$$

Non-idealities such as the residual offset (Fig. 4.7b) added to the DC signal of (4.8), or the electrical phase error (Fig. 4.7c) added by the transconductor to the AC signal of (4.6), lead to errors in the digitized value of ϕ_{ETF}. Apart from causing frequency error in DAFLL, via (4.2) these errors cause temperature sensing errors if the chip is used as a temperature sensor.

In the case of a residual offset current, I_{OS}, added to the demodulated DC signal (Fig. 4.7b), the modulator's feedback will still ensure that the average charge

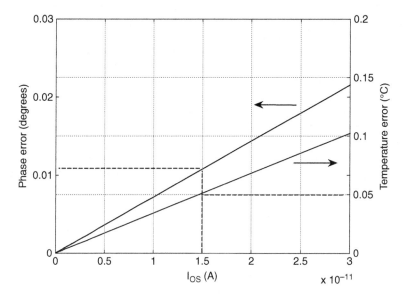

Fig. 4.8 Simulated phase error and the corresponding temperature measurement error as a function of residual offset current

accumulated by the integrator is approximately zero. However, a different steady state value μ' results:

$$\mu' \cdot \frac{1}{2} \cdot A \cdot \cos(\phi_{ETF} - \phi_0) + (1 - \mu') \cdot \frac{1}{2} \cdot A \cdot \cos(\phi_{ETF} - \phi_1) + I_{OS} \approx 0. \quad (4.12)$$

The deviation of μ' from the value expected from (4.9) leads to an error in the digitized ϕ_{ETF}. Figure 4.8 (left axis) illustrates the *simulated* phase error as a function of residual offset. The resulting temperature sensing error, calculated from (4.2), is also shown (right axis). A residual offset current of 15 pA, results in a phase error of approximately 0.01° and a temperature sensing error of 0.05°C. This simulation has been performed for the optimized ETF at room temperature, $f_{drive} = 85$ kHz, a heater power of 1 mW, and a transconductance of 300 μS.

Any residual DC offset current can be eliminated by chopping (see Fig. 4.5) the ETF and the entire front-end at a frequency f_{ch} (20 Hz), which is much lower than f_{drive} (85 kHz). The bit-stream output of the first chip was then decimated by a sinc[1] filter, with a length of N/f_{ch}, where N is an integer, so that its notches perfectly eliminate the ripple caused by the low-frequency chopping. This filter also limits the system noise bandwidth, thus suppressing most of the wide-band thermal noise of the ETF.

The other source of non-ideality is the electrical phase shift added to the transconductor's output current i_{gm}. The source of this phase shift is the finite bandwidth of all the circuitry between the ETF's output and the synchronous demodulator's input (Fig. 4.7c). This phase error ϕ_{error} is indistinguishable from ϕ_{ETF}. In case the first test chip is used as a temperature sensor, this directly translates to a temperature sensing error, and when used in a DAFLL, this phase error will translate to frequency error.

Adding ϕ_{error} to ϕ_{ETF} in (4.9), leads to a different steady-state value μ'':

$$\mu'' \cdot \frac{1}{2} \cdot A \cdot \cos(\phi_{ETF} + \phi_{error} - \phi_0)$$
$$+(1 - \mu'') \cdot \frac{1}{2} \cdot A \cdot \cos(\phi_{ETF} + \phi_{error} - \phi_1) \approx 0. \qquad (4.13)$$

From (4.13), it can be shown that for temperature sensing applications, a phase error of only $0.01°$ leads to an error of $0.05°C$. Assuming that the ETF is driven at 85 kHz, and that the transconductor behaves like a first order filter, this translates into a $-3dB$ bandwidth of more than 100 MHz.

4.3.2 PDΔΣM Circuit Design

As shown in Fig. 4.5, the PDΔΣM circuitry involves a transconductor with embedded chopper demodulator, which together with an integrating capacitor forms the modulator's passive integrator. Furthermore, it includes a differential-to-single ended amplifier and a bias circuitry. The design of these blocks will be described next.

4.3.2.1 Transconductor Design

As discussed in the system-level design section, the phase readout accuracy requires the transconductor of the PDΔΣM to be designed with a wide bandwidth. The thermal noise contribution of the transconductor should be less than that of the ETF. Finally, the residual offset current produced by the synchronous demodulator should be minimized. To satisfy these requirements, the transconductor was implemented as a gain-boosted folded-cascode amplifier, with PMOS input pair, and an embedded chopper demodulator (Fig. 4.9). In order to compare the performance of the bar ETF [3, 8] with the optimized ETF (Chap. 3), the transconductor employs two PMOS input pairs (M_{1-4}) connected in parallel. By multiplexing the tail current (60 μA) provided by the cascoded current source (M_{5-6}), any one of the two ETFs can be selected for use in the chip. Transistors M_{7-8} cascode the current sources M_{11-12} (each carrying 40 μA), while M_{13-14} (each carrying 10 μA) are cascoded by means of M_{9-10}. The offset current of the PMOS current sources (M_{13-14}) is chopped by means of the upper chopper.

The common-mode feedback (CMFB) circuit of the transconductor is formed by the two differential pairs made by $M_{15,16}$ and $M_{17,18}$ biased with current sources M_{19} and M_{20} together with the diode connected transistors M_{21} and M_{22} [10]. The gates of M_{16} and M_{17} are connected to the common mode reference voltage, V_{CM_ref}, which sets common mode level of the output terminals. The gate terminals

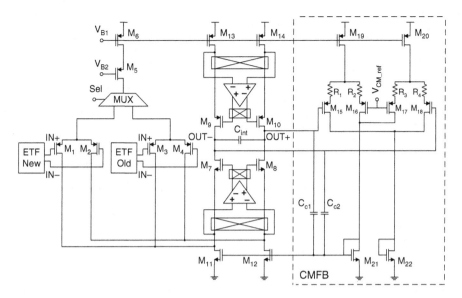

Fig. 4.9 Gain-boosted transconductor amplifier

of M_{15} and M_{18} sense negative and positive output terminals respectively. The sum of the output currents of M_{16} and M_{17} controls the output common mode level through the diode connected transistor M_{21} and the bottom current sources M_{11} and M_{12}. With the voltage of the positive and negative output terminals, the gates of M_{15} and M_{18} change at the opposite directions:

$$V_{G18} - V_{G17} = V_{G16} - V_{G15}. \qquad (4.14)$$

$$\frac{V_{G18} + V_{G15}}{2} = \frac{V_{G16} + V_{G17}}{2} = V_{CM\_ref}. \qquad (4.15)$$

In order to increase the linear range of operation for M_{15-18} to achieve a larger output swing, these devices are degenerated by resistors R_{1-4} (each 5 kΩ). Capacitors $C_{C1,2}$ (each 120 fF) provide frequency compensation for the CMFB loop in order to guarantee a minimum of 60° phase margin over the process corners and temperature.

Any electrical phase shift introduced between the input pair and the input of the synchronous demodulator gives rise to a phase error [see (4.13)]. The thermopile behaves like a distributed RC filter with a phase-shift of 0.14° at 85 kHz. To avoid increasing this phase-shift significantly, the transconductor's input capacitance (Fig. 4.10), $C_{par,gm}$, was designed to be small (50 fF) compared to the thermopile's total capacitance (600 fF), by making the input devices relatively small (36 μm/ 0.7 μm). The voltage across the input devices gives rise to a differential AC current $i_{sig,AC}$, which is then phase detected by the embedded chopper demodulator. Over temperature and process corners, simulations show that the minimum unity-gain

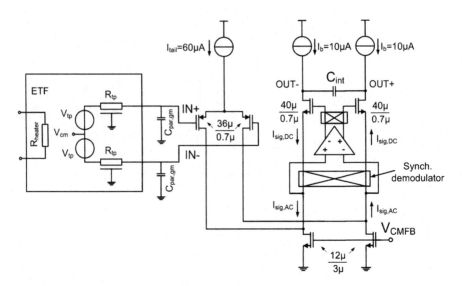

Fig. 4.10 Electrical filtering due to thermopile's parasitic capacitance, and due to the input pair's parasitic capacitance

BW of the corresponding transconductance is greater than 115 MHz. At $f_{drive} = 85$ kHz, this translates into a nominal phase shift of 0.01° in the AC current entering the synchronous demodulator, which is much smaller than the 0.14° of electrical phase shift introduced by the ETF itself. It should however be noted that the phase spread within a batch will be an order of magnitude smaller. Compared to the use of a preamp in [3, 8], the use of a single transconductor provides a significantly wider bandwidth for the AC current entering the demodulator – 115 MHz versus 25 MHz.

In a $\Sigma\Delta$ modulator based on a passive (g_m-C) integrator, the integrator leakage p, is determined by the DC output impedance of the transconductor R_{out}, the integrating capacitor C_{int}, and the sampling frequency f_S:

$$p = e^{-1/f_S R_{out} C_{int}}. \tag{4.16}$$

The width of the widest dead band Δx, in the modulator's DC characteristic normalized to its reference is given by [5, 11]:

$$\Delta x = \frac{1 - p}{1 + p}. \tag{4.17}$$

The amplifier's simulated DC gain is greater than 140 dB, which corresponds to a DC output impedance of more than 33 GΩ. With $C_{int} = 70$ pF, and a 2.67 kHz sampling frequency, this means that the dead bands associated with integrator leakage will be no wider than 0.004° in terms of ϕ_{ETF}. Via (4.2), this translates into a temperature measurement inaccuracy of less than 0.02°C if the chip is to be used as a TDC.

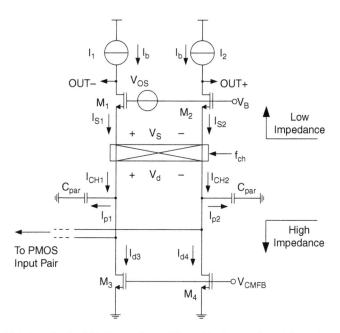

Fig. 4.11 Main branch of a folded-cascode amplifier with chopper demodulator

Simulations show that the transconductor has an input-referred thermal noise floor of 12.7 nV/√Hz (the ETF's noise level is at 18 nV/√Hz) and a $1/f$ noise corner of 50 kHz. Operating the chopper demodulator at 85 kHz ensures that most of the $1/f$ noise is modulated away from DC.

4.3.2.2   Residual Offset

Any offset current added to the current integrated by C_{int} will give rise to a phase measurement error [see (4.12)]. The chopper demodulator itself is a major contributor of such offset. This issue can be analyzed by considering the main branch of a classical PMOS-input folded-cascode amplifier shown in Fig. 4.11 (input pair not shown). Current sources I_1 and I_2, carrying a current I_b, represent the upper PMOS current sources, and the chopper demodulator is located at the sources of the cascode transistors M_{1-2} [12, 13]. In the presence of an offset V_{OS} between the NMOS cascode transistors, a net DC voltage V_S is established at the sources of the cascode transistors and across the chopper. The transient waveforms of the signals in the circuit of Fig. 4.11 are shown in Fig. 4.12. Due to the action of the chopper, a square-wave voltage V_d appears across the folding nodes of the amplifier. On the one hand, V_d charges and discharges the parasitic capacitors C_{par}, leading to an AC current $I_{p1}-I_{p2}$ [14], while on the other hand, it modulates the output currents of the bottom NMOS

Fig. 4.12 Transient waveforms of the circuit in Fig. 4.11

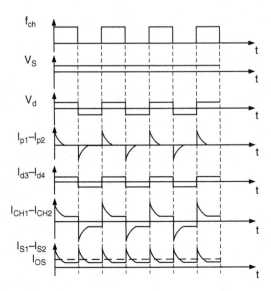

current sources, $M_{3\text{-}4}$ (with finite output impedances $R_{out3\text{-}4}$), leading to another AC current $I_{d3}\text{-}I_{d4}$. The sum of these two currents is an AC current, $I_{CH1}\text{-}I_{CH2}$, which is rectified by the chopper itself into an output current $I_{S1}\text{-}I_{S2}$. At a chopping frequency f_{ch}, the DC value of this current I_{OS} is:

$$I_{OS} = 4 \cdot f_{ch} \cdot C_{par} \cdot |v_d| + \frac{2 \cdot |v_d|}{R_{out3-4}}. \tag{4.18}$$

4.3.2.3 Gain-Boosting and Residual Offset Reduction

To minimize I_{OS}, the parasitic capacitances at the high-impedance input nodes of the chopper demodulator should be shielded from the DC voltage across its outputs. As shown in Fig. 4.13, a suitable location for the chopper is between the source terminals of the cascode transistors and the input terminals of the booster amplifiers. This way, the booster amplifiers [15, 16] will establish a virtual ground at the high-impedance, high-capacitance, folding nodes of the main amplifier. Therefore, the amplitude of the square-wave voltage V_d is reduced by the gain of the booster amplifier, and the amplitude of the offset current I_{OS}, as well. The output of the boosters must then also be chopped in order to maintain the correct feedback polarity. This technique has two advantages. Firstly, fixing the chopper's input nodes at virtual ground reduces the magnitude of I_{OS} by three orders of magnitude (from simulations), compared to the situation of non gain-boosted transconductor of

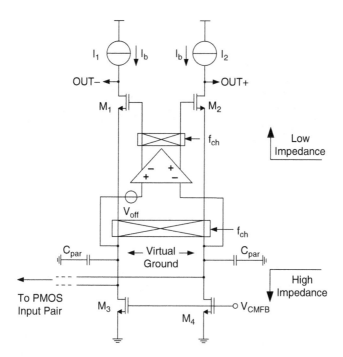

Fig. 4.13 Modified circuit turning the capacitive folding nodes to virtual grounds

Fig. 4.11. Secondly, chopping the booster's output means that the contribution of its offset (and $1/f$ noise) to the amplifier's output current is also chopped. Simulations show that, for a 10 mV worst-case offset between the cascode transistors and a 10% mismatch between the chopper switches, I_{OS} is about 150 pA. This is cancelled (Fig. 4.5) by chopping the entire front-end (and the ETF) at a lower frequency f_{ch} [17]. This should reduce the effect of residual offset to less than 0.01° in terms of ETF phase.

The requirements on the DC output impedance of the transconductor (to limit the dead bands) mean that the booster amplifiers have to have a DC gain in excess of 60 dB. To obtain this, as well as the highest possible common mode rejection ratio, they were implemented as fully differential folded-cascode amplifiers. Figure 4.14 illustrates the booster amplifier of the NMOS cascode devices of the main amplifier (Fig. 4.9). A similar topology is used for the booster amplifier of the PMOS cascode devices. Input common-mode regulation is used to set the common-mode level of the booster's input terminals [10]. The input pair (M_1 and M_2) is provided with two extra transistors (M_3 and M_4) in a common-source configuration, whose gates are connected to the input common mode reference, V_{cm_in}. Due to the feedback between the input and output of the booster amplifier via the main cascode transistors, the output common-mode voltage is regulated in such a way that the input common-mode voltage is equal to V_{cm_in} [10]. The booster amplifiers each consume 30 µA from the 5 V supply.

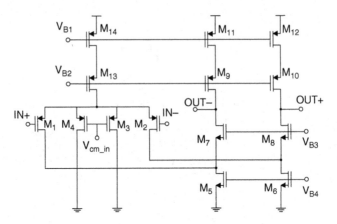

Fig. 4.14 The booster amplifier for the NMOS cascode transistors with input common-mode regulation

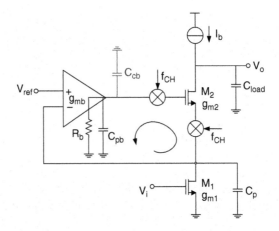

Fig. 4.15 Gain-boosting loop around a cascode transistor including choppers and parasitic capacitances (single-ended half circuit)

It should be noted that the gain-boosting loop around the cascode transistors includes chopper switches at the source of the cascodes as well as the output of the booster amplifier (see Figs. 4.9 and 4.13). The switching action occurring in the loop requires that the loop has enough phase-margin to guarantee its stability during the switching transients. This can be better understood if the simplified half-circuit shown in Fig. 4.15 is considered. Ignoring the parasitics due to the chopper switches and without the compensation capacitor C_{cb} the loop has two poles (see the dotted line open loop gain in Fig. 4.16). This is due to the parasitic capacitance C_p at the drain of M_1 and the output capacitance of the booster C_{pb}.

This characteristic will lead to instability since the unity-gain frequency is higher than the non-dominant pole. By adding a compensation capacitor C_{cb} to

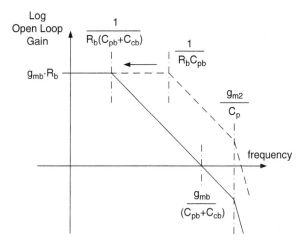

Fig. 4.16 Open-loop gain of the gain boosted cascode circuit in Fig. 4.15

the output of the booster the characteristic will be compensated to the solid line in Fig. 4.16. Thus the stability condition of the loop requires that:

$$\frac{g_{m2}}{C_p} \succ \frac{g_{mb}}{C_{cb}}. \tag{4.19}$$

To guarantee the stability of the gain-boosting loops in the circuit of Fig. 4.9, load capacitances of 900fF are added between the output terminals of the boosters and ground.

4.3.2.4 Differential-to-Single-Ended Amplifier

Since the signal swing across C_{int} is too small to drive an off-chip latch, it is buffered by a differential-to-single-ended amplifier with a DC gain of 40 dB and a rail-to-rail output (Fig. 4.17). The amplifier consists of a degenerated ($R_1 = R_2 = 10$ kΩ) differential input pair and three current mirrors. The NMOS current mirrors have a gain of 2, while the PMOS ones have a gain of 1. The amplifier has one dominant pole (due to the parasitic capacitance of the pad it drives), which is located at its output terminal.

4.3.2.5 Bias Circuit

A constant-g_m bias circuit [18] produces the biasing current for the transconductor as well as the differential-to-single-ended amplifier (see Fig. 4.18). This circuit produces a bias current for the input pair of the transconductor, such that its transconductance is to first order only determined by the resistor R. The current

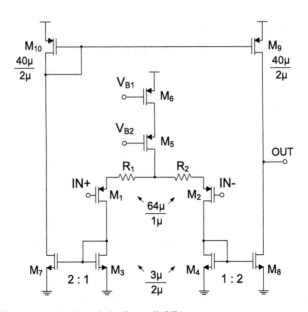

Fig. 4.17 Differential-to-single-ended rail-to-rail OTA

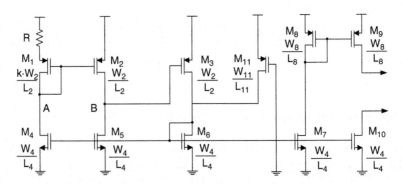

Fig. 4.18 Constant-g_m bias generator circuit

produced in transistor M_6 is mirrored to M_7 and M_8 and eventually to the drains of M_9 and M_{10} from which the various circuit blocks are biased.

Considering that the PMOS input pair of the transconductor is biased at strong inversion, the overdrive voltage, V_{od}, of each device carrying a drain current of I_d is given by:

$$|V_{od}| = |V_{gs} - V_{TH,P}| = \sqrt{\frac{2}{\mu_P C_{OX}} \frac{L}{W} I_d}. \qquad (4.20)$$

where $V_{TH,P}$ is the threshold voltage and μ_P is the mobility of the PMOS device, while C_{OX} is the gate oxide unit capacitance. Considering the voltage across resistor R to be V_R and the gate-source voltages of transistors M_1 and M_2 denoted by V_{gs1} and V_{gs2}:

$$V_R = |V_{gs2}| - |V_{gs1}|$$

$$= \sqrt{\frac{2}{\mu_P C_{OX}} \frac{L_2}{W_2} I_{d2}} - V_{TH2} - \sqrt{\frac{2}{\mu_P C_{OX}} \frac{L_1}{W_1} I_{d1}} + V_{TH1}. \qquad (4.21)$$

On the other hand, I_{d1} and I_{d2} are equal. This is because $V_{gs4} = V_{gs5}$ and $V_A = V_B$ (minimizing the short channel effects in M_4 and M_5). The latter is due to the feedback provided by the current mirror including M_{4-6} and transistor M_3 ensuring that $V_{gs3} = V_{gs2}$ and as a result $V_A = V_B$. Assuming equal threshold voltages for M_1 and M_2, (4.21) can be simplified to:

$$V_R \approx \left(\sqrt{\frac{L_2}{W_2}} - \sqrt{\frac{L_1}{W_1}} \right) \sqrt{\frac{2}{\mu_P C_{OX}} I_d}. \qquad (4.22)$$

where $I_d = I_{d1} = I_{d2}$. Substituting $V_R = R \cdot I_d$ and considering that $L_1 = L_2$ and $W_1 = k \cdot W_2$:

$$I_d = \frac{1}{R^2} \frac{2}{\mu_P C_{OX}} \frac{L_1}{W_1} \left(1 - \frac{1}{\sqrt{k}} \right)^2. \qquad (4.23)$$

this current is mirrored and used to bias the PMOS devices in the input pair of the transconductor. If their current densities, and hence their region of operation is kept the same as that of M_2, their transconductance will be to first order only determined by resistor R and therefore not a function of temperature:

$$G_m = \sqrt{2 \mu_P C_{OX} \frac{W}{L} I_d} = \frac{2}{R} \left(1 - \frac{1}{\sqrt{k}} \right). \qquad (4.24)$$

Furthermore, transistor M_{11} is a very long device ($W/L = 1/300$) that produces a startup current path from V_{DD} to M_6 allowing for the gate voltage of the NMOS devices to be raised, which then turns on the PMOS devices as well.

4.3.3 First Chip Experimental Results

As described in Sect. 4.2.3, besides the phase domain $\Delta\Sigma$ modulator, the first test chip included two ETFs. These were the bar ETF used previously in [3, 8], and the

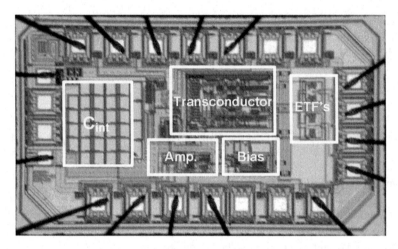

Fig. 4.19 Micrograph of the first test chip including ETF's and the phase-domain delta-sigma modulator

optimized ETF. This chip is realized in a standard 0.7 μm CMOS technology and has an area of 2.3 mm² (Fig. 4.19) and is packaged in a ceramic DIL package.

The selected ETF and the ΔΣ modulator each consume 2.5 mW from a 5 V supply. The timing signals were generated in an FPGA and derived from a 16 MHz crystal oscillator. The sampling rate of the modulator was 2.67 kHz, and the low frequency chopper was driven at 20 Hz. The modulator's output was decimated by a 14-bit counter, which acts as a 1st-order sinc filter and limits the system bandwidth to 0.16 Hz. A sinc¹ filter was chosen over a more complex sinc² filter, because the former actually achieves slightly more resolution. This is because the modulator's resolution is mainly limited by the ETF's thermal noise, which, for the same filter length, is more effectively suppressed by the narrower noise bandwidth of a sinc¹ filter. Figure 4.20 shows the output spectrum of the phase domain ΔΣ modulator at room temperature before and after decimation. The tone at 20 Hz is due to use of the low-frequency chopping, and is completely suppressed by the decimation filter. The variance of the measured noise in the decimated bit-stream was 0.006° (rms) in terms of the ETF phase, which is better than 13 bits. This noise corresponds to a temperature-sensing resolution of 0.03°C (rms).

To measure the accuracy of the devices, they were mounted in good thermal contact with a large aluminum block, and their temperature was then measured by a PT-100 temperature sensor. The implementation of an on-chip multiplexer, meant that TD temperature sensors based on both the bar ETF and the optimized ETF could be characterized. The averages of the measured phase-versus-temperature characteristics of 16 devices are shown in Fig. 4.21, as well as the simulated characteristics. These characteristics were obtained by inverting (4.10) and then performing an 8th order least-squares polynomial fit on the measured data, i.e. the output of the PT-100 sensor and the decimated output of the chips. Due to their different geometries, the slope of their phase characteristics is significantly

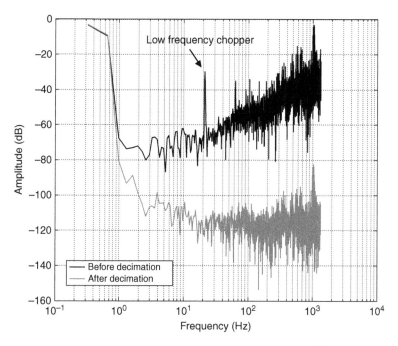

Fig. 4.20 Measured output spectrum of the TDC at room temperature before and after decimation (8192-point FFT, Hanning window)

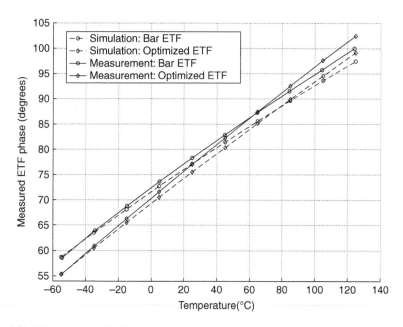

Fig. 4.21 Measured and simulated phase characteristics of TDCs based on both ETFs

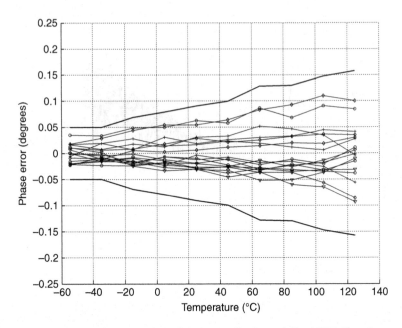

Fig. 4.22 Measured deviation of each TDC's phase (using the optimized ETF) from the average phase-temperature characteristic (*bold line*: 3σ error)

different. From these results, the maximum value of the sensitivity function $S^T_{\phi_{ETF}}$ of the bar ETF is 4.7°C/degrees, while that of the optimized ETF is 4.1°C/degrees. The latter is slightly smaller than the simulated value by (4.2).

The average characteristics were then used to translate the decimated output of *each* chip into an absolute temperature value. For the optimized ETF, the phase deviation from the average characteristic is shown in Fig. 4.22 (16 devices). This shows a measured phase characteristic of 0.15° (3σ). The corresponding temperature deviation is shown in Fig. 4.23. It can be seen that a TDC based on the optimized ETF achieves an untrimmed inaccuracy of about ±0.7°C (3σ) over the military range (−55°C to 125°C), while a TDC based on the bar ETF only achieves an inaccuracy of ±0.8°C (3σ).

4.3.4 Conclusions from the First Test Chip

These experimental results show that the optimized ETF achieves better phase accuracy compared to the bar ETF. They also conclude that an ETF's phase shift can be digitized with a resolution better than 13 bits and accuracy better than 0.15°, showing a feasible platform for the implementation of the DAFLL in the second test chip. As discussed earlier in this chapter, phase accuracy of about 0.15° should

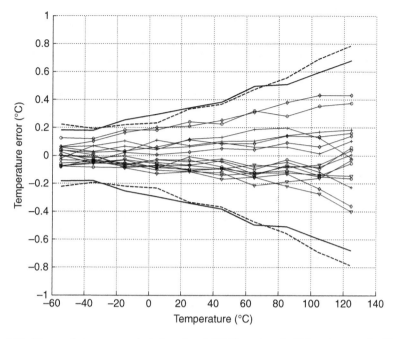

Fig. 4.23 Measured temperature deviation of each TDC's output (*bold line*: 3σ error for TDC's based on the optimized ETF, *dotted line*: 3σ error for TDC's based on the bar ETF)

allow for frequency accuracies better than 0.3%. Therefore, in the next phase a combination of the optimized ETF and the PDΔΣM were combined with a digitally-controlled oscillator (DCO) to build a DAFLL.

4.4   Second Test Chip

The second test chip builds a complete DAFLL based on the optimized ETF and the phase-domain delta sigma modulator (PDΔΣM), whose performances were demonstrated experimentally by the first test chip. The second test chip adds a digitally-controlled oscillator (DCO) to the circuitry developed in the first chip. The DCO's system-level specifications were described earlier in Sect. 4.2.2.

Based on the required DCO specifications, a system-level design of the DCO will be provided. Then, a more complete system-level block diagram of the DAFLL (compared to the block-diagram shown in Fig. 4.3) will be provided. This block-diagram involves the system-level diagrams of the PDΔΣM (described in previous sub-section) and the DCO. Results of time-domain behavioral simulations of this system will be provided, which demonstrate the step response and stability of the DAFLL system.

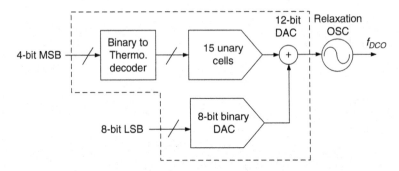

Fig. 4.24 The 12-bit DCO made of a segmented DAC and a relaxation oscillator

After these system-level discussions, the circuit design of the DCO (as the only new analog block in the second test chip) will be provided. This will then be followed by experimental results from the second test chip.

4.4.1   DCO System-Level Architecture

In principle, a DCO can be based on any type of oscillator. At RF frequencies, DCOs based on LC oscillators with a digitally-controlled bank of capacitors have been proposed [19]. With the range of frequencies produced by the DAFLL and the capabilities of the 0.7 µm CMOS technology used, the DCO was realized as a 12-bit current steering DAC whose output current tunes the output frequency of a relaxation oscillator (Fig. 4.24). The 12-bit requirement of the DCO resolution has been discussed in Sect. 4.2.2.

To guarantee the monotonic behavior required to ensure stability of the DAFLL, the DAC has a segmented architecture. The four MSBs are fed via a binary-to-thermometer decoder to a unary DAC made of 15 equal elements, while the eight LSBs drive a binary-weighted DAC. The matching requirements determining the monotonicity of the DAC are now relaxed to that of an 8-bit binary DAC, which can be achieved by proper sizing and careful layout. The sum of the output currents from both segments is fed to the oscillator.

4.4.2   Complete DAFLL System-Level Simulations

A complete system-level block diagram of the second test chip is shown in Fig. 4.25. Based on the system shown in this diagram, time-domain and frequency-domain behavioral Matlab simulations were performed on the DAFLL. This was mainly to study the response of the DAFLL to a step applied to the DCO input, as well as to study the output spectra of the PD$\Delta\Sigma$M and the digital integrator

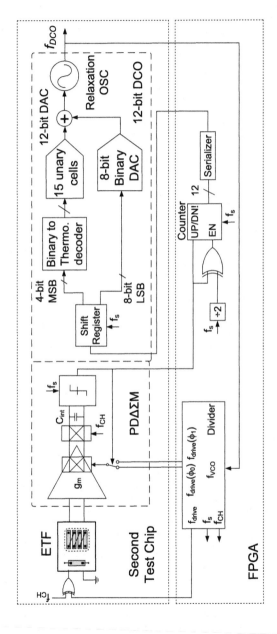

Fig. 4.25 The complete system-level block diagram of the DAFLL (the second test chip)

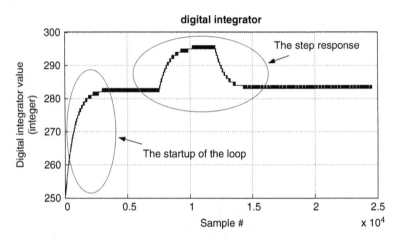

Fig. 4.26 Time-domain behaviorally simulated step response of the DAFLL

(up/down counter). The details on the time-domain simulation method can be found in Appendix A. For the time-domain simulation of the ETF output signal, the RC network described in Chap. 3 has been adopted (see Fig. 3.15).

In the behavioral Matlab simulations, the ETF is driven at a frequency $16\times$ lower than that of the DCO. Its output signal is fed to a time domain model of the PD$\Delta\Sigma$M with phase references that are $\pm45°$ phase shifted in reference to the ETF heater drive. The ETF heater dissipates 5 mW of power and its thermopile has a sensitivity of 0.5 mV/K. The PD$\Delta\Sigma$M's loop filter (g_m-C combination) has a unity gain frequency of 300 kHz and the DAC incorporates 12 bits with a 4 V reference voltage. The VCO has a sensitivity of 3.6 MHz/V and the whole DAFLL is sampled at a sampling frequency which is $2\times$ lower than the ETF's heater drive frequency.

After the loop stabilizes to a DCO frequency corresponding to a 90° ETF phase shift, a step voltage of 100 mV is applied to the VCO input. The response of the loop can be monitored through the digital integrator's output (see Fig. 4.26), which shows a first order settling. This is the same behavior seen in the analog electrothermal FLL showing that the digitally-assisted loop also remains first-order and therefore unconditionally stable. This is because the sampling rate of the loop is much larger than the loop bandwidth [2] (in these simulations, the sampling rate is >50 kHz and the loop bandwidth is about 1 Hz).

The time-domain simulation shows the limit cycle behavior of the DAFLL, where at the steady state of the loop the digital integrator's output toggles between two LSB's. The amplitude of the frequency variation at the output caused by this idle tone will be determined by the DCO's minimum step size. Furthermore, the output spectrum of the PD$\Delta\Sigma$M and the digital integrator can be seen in Fig. 4.27. The former has a 20 dB/decade noise shaping and the latter shows sufficient filtering of the PD$\Delta\Sigma$M's quantization noise.

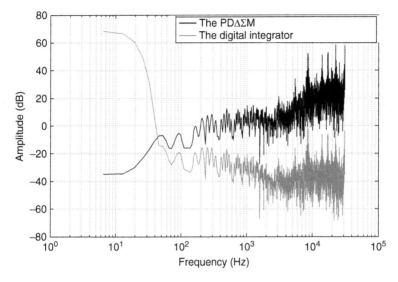

Fig. 4.27   Output spectra of the PD$\Delta\Sigma$M and the digital integrator of the DAFLL (24,576 samples, Kaiser Window, N = 20)

4.4.3   DCO Circuit Design

The DCO is the only new analog block added to the first test chip, and so, the circuit design of the second chip, only involves the design of the DCO. This includes a 12-bit DAC and a relaxation oscillator, which are described in the following subsections.

4.4.3.1   Relaxation Oscillator

The oscillator [20, 21], and its transient waveforms are shown in Fig. 4.28. Depending on the state of the latch, either capacitor C_1 or C_2 is charged to V_{dd} by transistor M_1 or M_3 respectively, while the other capacitor is gradually discharged by I_{ref} via M_2 or M_4. I_{ref} is provided by the cascode current mirror consisting of M_{5-8}. When the voltage across the capacitors reaches V_{ref}, comparators $Comp_{1,2}$ feed the latch with the appropriate set or reset pulses, which then toggles after a delay, $t_d \sim 100$ ns. The period of oscillation is:

$$T_{OSC} = \frac{2 \cdot C \cdot (V_{dd} - V_{ref})}{I_{ref}} + t_d. \tag{4.25}$$

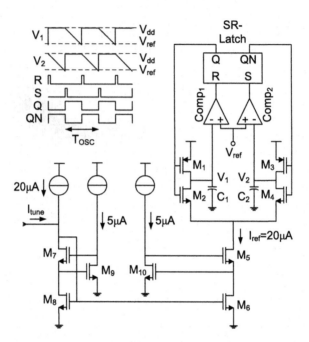

Fig. 4.28 Circuit and timing diagram of the relaxation oscillator

in which $C = C_1 = C_2 = 1$ pF. For $I_{ref} = 20$ μA, $V_{dd} = 5$ V, and $V_{ref} = 2.25$ V, the center frequency is 2.6 MHz. To vary this frequency, the reference current is varied by pushing or pulling a current I_{tune} into the current mirror (drain of M_7). This current is provided by the DAC, and was chosen to be between ± 18 μA, which corresponds to a tuning range from 350 kHz to 4 MHz. This range is wide enough to accommodate the temperature dependent variation of the FLL's frequency as well as the spread of the DCO itself. The cascode transistors M_5 and M_7 are gain-boosted by transistors M_9 and M_{10}. This ensures that the mirror ratio is well defined despite the voltage excursions on $C_{1,2}$ and the variation of I_{tune}.

The comparators $Comp_{1,2}$ (Fig. 4.29) consist of positive-feedback latches with 50 mV of hysteresis, preceded by pre-amplifier made by $M_{6,7}$ differential pair and load transistors $M_{8,9}$ (total gain of ≈ 5) in order to minimize the effect of their "kickback." Hysteresis, required to reject the noise on the ramp voltage produced on the oscillator's capacitors, is produced in the latch by making the $\beta_n = \mu_n C_{ox}(W_n/L_n)$ of $M_{10,13}$ different from that of $M_{11,12}$ (50%). In this oscillator, the main cause of jitter is the input-referred noise of the comparators [22, 23] and the $1/f$ noise of reference current. With a supply current of 70 μA, over process corners and temperature, the maximum input-referred noise of the comparator is 140 μV (rms), which translates to 50 ppm of period jitter at an oscillation frequency of 1.6 MHz [22].

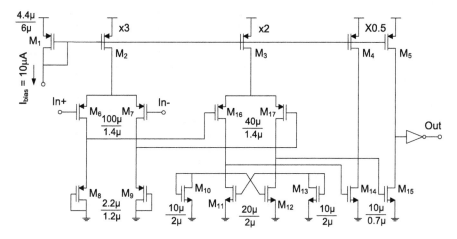

Fig. 4.29 Schematic of the comparator used in the relaxation oscillator

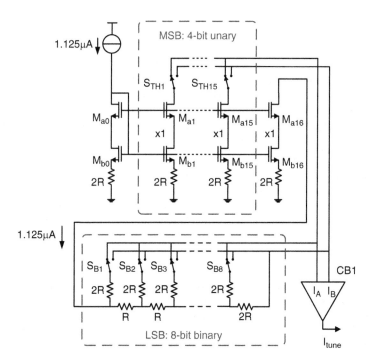

Fig. 4.30 Circuit diagram of the segmented current steering DAC

4.4.3.2   12-Bit Current-Steering DAC

The 4-bit MSB part of the segmented DAC (Fig. 4.30) is implemented by means of 15 equal current sources, while the 8-bit LSB part is made by means of an R–2R ladder network [24, 25]. The DAC's reference current is 1.125 µA, which is copied

Fig. 4.31 Schematic
of the current buffer CB1

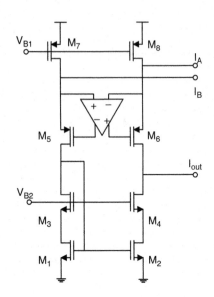

15 times by current sources $M_{a,b1-15}$, which are degenerated by resistors 2R. The 15 switches S_{TH1-15} are driven by a thermometer code representation of the 4 MSBs. An extra copy of the reference current is generated by $M_{a,b16}$ to provide a 1.125 µA reference current for the R–2R network (R = 80 kΩ). This network divides the current into binary-weighted fractions, which are selected by the $M_{a,b1-15}$, which are degenerated by resistors 2R. The 15 switches S_{TH1-15} are driven by a thermometer code representation of the 4 MSBs. An extra 8 switches S_{B1-8} driven by the 8 LSBs. The output currents of the unary and binary sections are added in a differential-to-single ended current buffer, CB_1 (Fig. 4.31), and thus form the oscillator's tuning current I_{tune}. The output current of the resistor ladder enters the current buffer at the virtual ground points provided by the gain booster of the cascode transistors $M_{5,6}$. The current mirror made of M_{1-4} then translates this current into a single-ended current.

For a monotonic characteristic, the matching between the reference current sources $M_{a,b,16}$ and the unary sources should be better than 8-bits (0.4%). Therefore, the current sources were carefully laid out in a common-centroid manner. The input-referred offset of CB_1 produces an offset current on the R–2R network, which should be less than the current corresponding to one LSB. This translates to a worst-case input referred offset of 2 mV, which was achieved by proper sizing and careful layout.

4.4.4   Experimental Results with the Second Test Chip

For experiments with the second test chip, the critical components of the digitally-assisted electrothermal FLL (DAFLL) were realized in a standard 0.7 µm CMOS

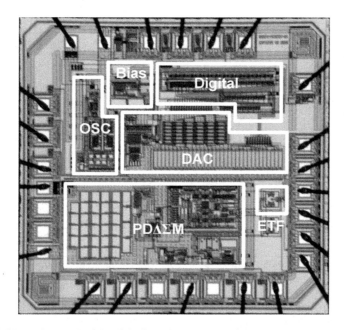

Fig. 4.32 Photomicrograph of the digitally-assisted electrothermal FLL

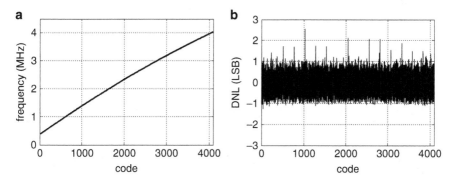

Fig. 4.33 DCO characteristic, (**a**), the DAC's DNL, (**b**)

process [4]. The test chip (Fig. 4.32) has a die area of 4 mm^2. The optimized ETF dissipates 2.5 mW, while the PD$\Delta\Sigma$M and the 12-bit DCO together dissipate 5 mW from a 5 V supply. For flexibility, the frequency divider and the up/down counter were realized in an FPGA.

The measured characteristic of the DCO versus input code is shown in Fig. 4.33a. Its tuning range is sufficiently large, and its non-linear characteristic will be compensated for by the FLL. The measured DNL of the DAC is shown in

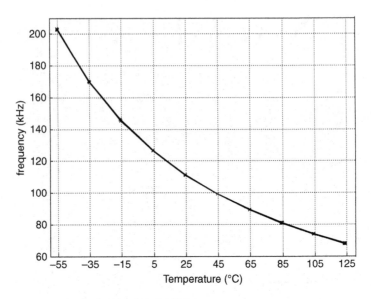

Fig. 4.34 Measured characteristic of the DAFLL over temperature

Fig. 4.33b. Although the DAC is not monotonic (DNL <-1 LSB), this did not cause problems in practice, because the loop is effectively dithered by the ETF's thermal noise (about 2 LSBs p-p). The DAC's LSB corresponds to a DCO step of 890 Hz. The sampling rate of the PD$\Delta\Sigma$M and the DCO was set to $f_{drive}/32$, which corresponds to a noise-bandwidth of 0.4 Hz (at room temperature). The jitter in f_{drive} $(= f_{VCO}/16)$ was about 500 ps (rms), which corresponds to a temperature sensing resolution of 0.015°C (rms). Measurements on 16 samples from one batch show the expected $1/T^{1.8}$ dependency of f_{drive} (Fig. 4.34). The spread in the FLL's output frequency is about $\pm 0.3\%$ (3σ) (Fig. 4.35a) from -55°C to 125°C. This corresponds to a temperature measurement inaccuracy of about 0.7°C (3σ) (Fig. 4.35b). This level of accuracy is comparable to that of the analog electrothermal FLL's reported in [3, 26]. However, those required the use of large external capacitors.

4.5   Measuring the Effective Thermal-Diffusivity of CMOS Chips Using a DAFLL

The proposed electrothermal DAFLL can be used in order to determine the effective value of the thermal diffusivity, D_{eff}, of a CMOS chip. D_{eff} is a parameter that describes the rate at which heat diffuses through a chip, and hence its knowledge is essential for the thermal management of systems on chip and the design of thermal sensors. By embedding an electrothermal filter (ETF) in a frequency-locked-loop

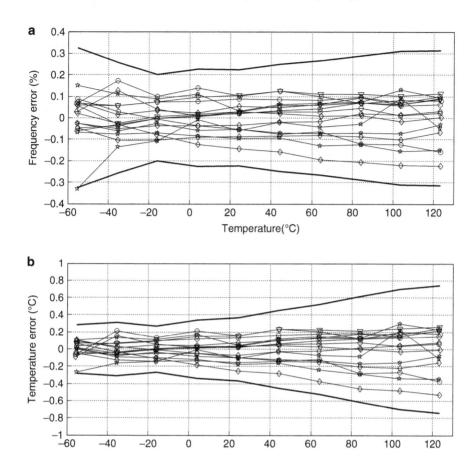

Fig. 4.35 Output frequency error of DAFLL measured for 16 samples (*bold lines*: ±3σ borders), (**a**), the equivalent temperature sensing inaccuracy (*bold lines*: ±3σ borders), (**b**)

(FLL), its phase response, which is determined by its (fixed) geometry and D_{eff}, can be measured. D_{eff} can then be accurately determined from the measured phase response [27].

4.5.1 The Essence of Measuring D_{eff}

The trend towards smaller and smaller devices in modern CMOS IC technology, has led to the realization of complex integrated systems on chip, e.g. advanced microprocessors. In such systems, however, the associated increase in the dissipated power density means that activity-dependent hot spots may be formed on the surface of the chip. The resulting peak junction temperatures affect system reliability, and so require various forms of thermal management [28].

Successful thermal management relies on good models of the rate at which heat diffuses from the heat-dissipating devices into the rest of the chip. For a given CMOS process, this will be a complex function of the thickness and thermal properties of the field oxide, the epitaxial layer and the underlying substrate. However, for practical purposes, the rate of heat diffusion can be modeled by a single parameter: the effective thermal diffusivity of the die, D_{eff} [29].

Given the thickness of the various layers of a chip, and their thermal properties, D_{eff} can be determined by numerical modeling [29, 32]. However, these properties are usually not precisely known. In fact, even the reported values of the thermal diffusivity of pure silicon vary considerably – from 0.758 to 0.960 cm^2/s at room temperature – probably due to differences in measurement methods [30]. In consequence, D_{eff} can best be determined by measurements on electrothermal filters realized in processed chips. The phase shift of an ETF, ϕ_{ETF}, is a function of its geometry (fixed by its layout) and D_{eff} [3].

By using a CMOS ETF as the frequency-determining element of a frequency-locked loop (FLL) [3, 26], the frequency corresponding to a given phase shift ϕ_{ETF} and, hence, the ETF's phase response can be accurately determined. Since this response is only a function of the filter's (fixed) geometry and D_{eff}, the latter can then be accurately determined by using it as a curve-fitting parameter to fit the measured phase response with that predicted by numerical modeling.

Since silicon is a good thermal conductor, an ETF has a sub-millivolt output signal when driven by a few milliwatts of heater power. Therefore, it is preferable to characterize the phase response of an ETF with an on-chip FLL. This approach avoids the extra phase shift that would otherwise be incurred due to the interaction between the thermopile's resistance and the parasitic capacitance associated with the connections to an off-chip FLL. This approach is also a reasonably generic method of determining D_{eff}, since an ETF and an on-chip FLL can be implemented in any CMOS process.

4.5.2   Thermal Diffusivity Measurement Using CMOS ETFs

A semi-analytical model of the ETF needs to be used for the thermal diffusivity measurements. As described in Sect. 3.2, a simplified point-heater and point-temperature model described by (3.8) describes the frequency domain transfer function relating the dissipated heat in the heater to the small temperature variations detected by the thermopile. As described by (3.12) in Sect. 3.4, this model can be expanded to a practical ETF structure, which is more complex than the point source model.

With the help of the semi-analytical model, ϕ_{ETF} can be determined as a function of the filter's excitation frequency, its geometry, and D_{eff}. The model assumes that

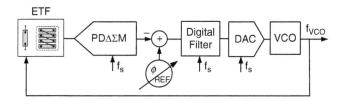

Fig. 4.36 Concept of using an electrothermal FLL for D_{eff} measurement

the substrate is homogenous and has a constant thermal diffusivity. The effective thermal diffusivity of the substrate can then be determined by using D_{eff} as a fitting parameter to match the phase response predicted by the model with the measured phase response of the ETF.

4.5.3   An Electrothermal FLL as a Test Vehicle in Measuring D_{eff}

In order to determine the value of D_{eff}, the ETF's phase response needs to be accurately characterized. This can be done by using the ETF as the frequency-determining element of a frequency-locked loop (FLL), whose phase set-point can be digitally adjusted.

A conceptual block-diagram of the proposed method including the FLL in a D_{eff} measurement is shown in Fig. 4.36. The FLL is the same one implemented in the second test chip, with a slight modification in the logic circuitry in the FPGA. The digital output of the PD$\Delta\Sigma$M, representing ETF's phase shift is compared with a digitally programmable phase set-point, ϕ_{REF}. The resulting error signal is integrated by the digital filter and fed back, via a DAC, to the VCO. The feedback forces the VCO to operate at $f_{VCO} = f_{REF}$, where f_{REF} is the excitation frequency at which $\phi_{ETF} = \phi_{REF}$. The PD$\Delta\Sigma$M, the digital filter, and the DAC are all sampled at the same frequency f_s, which is a sub-multiple of f_{VCO}. By varying ϕ_{REF} digitally and measuring f_{REF}, the ETF's phase response can be extracted.

4.5.4   Experimental Results

Based on the concept proposed in the previous subsection, experiments were performed on the chips shown in Fig. 4.32. The phase response of the ETF was measured for four devices at three temperatures: $-55°C$, $27°C$ and $125°C$ (Fig. 4.37). This was done by sweeping the phase set-point ϕ_{REF} of the electrothermal FLL digitally and measuring the corresponding f_{REF}, while the chip's temperature was controlled in a temperature-controlled chamber.

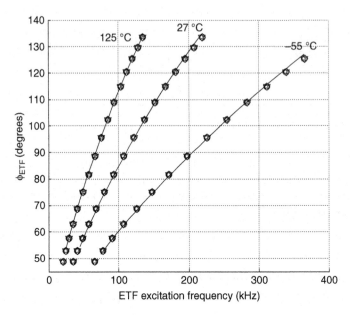

Fig. 4.37 Measured phase-frequency characteristic of the ETF for four devices (data points) versus the simulated characteristic of ETF with fitted values of D_{eff} (*bold lines*)

By using D_{eff} as a fitting parameter, the characteristic predicted by the analytical thermal model of the ETF (Fig. 4.37, bold lines) was fitted to the measured data. The fit shows a very good agreement between the measurements and the model. The corresponding values of D_{eff} that fitted the simulation to measurement were 1.405, 0.755, and 0.495 cm^2/s at $-55°$C, $27°$C, and $125°$C, respectively. These values show the $T^{-1.8}$ dependency expected from literature (Fig. 4.38). The room temperature value of D_{eff} also agrees well with the results of a boundary-element-method analysis of the ETF and its substrate [29]. It should be noted that the room temperature value of D_{eff} is considerably lower than that of bulk silicon. This is mainly due to the lower thermal conductivity of the heavily doped substrate [31].

4.6   Conclusions

An electrothermal frequency-locked loop (FLL) locks the output frequency of a VCO to the phase shift of an electrothermal filter (ETF). An ETF's phase shift is a function of its geometry, determined by lithography, and the temperature-dependent thermal-diffusivity of silicon, D. The latter is process insensitive for the low doping levels of IC-grade silicon, and so the accuracy of an ETF is limited by lithography. This means that in an FLL, the VCO's own process spread and temperature drift will no more affect its output frequency.

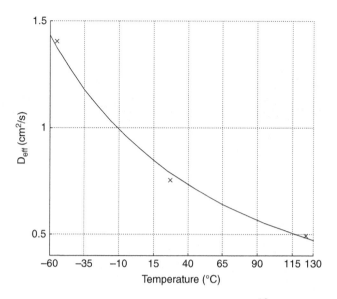

Fig. 4.38 Measured value of D_{eff} (data points) compared to the $T^{-1.8}$ temperature dependency predicted by literature (*bold line*)

The main drawback of previous all-analog FLL's was that they needed an off-chip capacitor to reduce the loop noise bandwidth and to filter the large low frequency ripple associated with chopping. This does not lend itself to a solution amenable for CMOS integration. Therefore, a digitally-assisted electrothermal FLL (DAFLL) is proposed in which the required narrow noise bandwidth was implemented by a digital filter. Such loop achieved an output frequency accurate to $\pm 0.3\%$ from $-55°C$ to $125°C$. The output frequency of the loop followed the same $T^{-1.8}$ trend associated with the temperature dependence of D, measured for the prior analog FLL's.

Such a DAFLL can be used to measure the effective thermal diffusivity constant of a CMOS chip. The loop was used to measure an ETF's phase frequency characteristic over temperature. By using the effective value of D as a curve fitting parameter, the measured characteristic could be fitted to that given by an analytical model, and hence its effective value could be determined. These values are 1.405, 0.755, and 0.495 cm^2/s at $-55°C$, $27°C$, and $125°C$, respectively. The room temperature value of D_{eff} is considerably lower than that of bulk silicon reported in the literature, which is probably due to the lower thermal conductivity of the heavily doped substrate

The digitally-assisted electrothermal FLL provides the basis for the generation of an accurate and stable on-chip frequency reference. It will be shown in the next chapter that this can be achieved by temperature-compensating the loop.

References

1. Murmann B (2006) Digitally assisted analog circuits. IEEE Micro 26(2):38–47
2. Staszewski RB, Balsara PT (2006) All-digital frequency synthesizer in deep-submicron CMOS. Wiley, Hoboken
3. Makinwa KAA, Snoeij MF (2006) A CMOS temperature-to-frequency converter with an inaccuracy of less than ±0.5 °C (3σ) from −40 °C to 105 °C. IEEE J Solid-State Circ 41(12):2992–2997
4. Kashmiri SM, Makinwa KAA (2009) A digitally-assisted electrothermal frequency-locked loop. In: Proceedings of the 35th ESSCIRC, Athens, Greece, pp 296–299
5. Schreier R, Temes GC (2005) Understanding delta-sigma data converters. Wiley, Hoboken/Chichester
6. Breems LJ, Huijsing JH (2001) Continuous time sigma delta modulation for A/D conversion in radio receivers. Kluwer Academic Publishers, Dordrecht, The Netherlands, Springer
7. Kashmiri SM et al (2009) A temperature-to-digital converter based on an optimized electrothermal filter. IEEE J Solid-State Circ 44(7):2026–2035
8. van Vroonhoven CPL, Makinwa KAA (2008) A CMOS temperature-to-digital converter with an inaccuracy of ±0.5 °C (3σ) from −55 to 125 °C. In: IEEE ISSCC Dig. Tech. Papers, San Francisco, CA, February 2008, pp 576–577
9. van Vroonhoven CPL et al (2010) A thermal-diffusivity-based temperature sensor with an untrimmed inaccuracy of ±0.2 °C (3σ) from −55 °C to 125 °C. In: IEEE ISSCC Dig. Tech. Papers, San Francisco, CA, pp 314–315
10. Huijsing JH (2001) Operational amplifiers theory and design. Kluwer, Boston. ISBN 0-7923-7284-0
11. Feely O, Chua LO (1991) The effect of integrator leak in Σ-Δ modulation. IEEE Trans Circuit Syst 38(11):1293–1305
12. Denison T et al (2007) A 2 μW 100 nV/rtHz chopper-stabilized instrumentation amplifier for chronic measurement of neural field potentials. IEEE J Solid-State Circ 42(12):2934–2945
13. Sanduleanu M et al (1998) A low noise, low residual offset, chopped amplifier for mixed level applications. In: Proceedings of the IEEE international conference on electronics, circuits and systems, Lisboa, Portugal, vol 2, pp 333–336
14. Witte JF, Makinwa KAA, Huijsing JH (2007) A CMOS chopper offset-stabilized opamp. IEEE J Solid-State Circ 42(7):1529–1535
15. Bult K, Geelen GJGM (1990) A fast-settling CMOS op amp for SC circuits with 90-dB DC gain. IEEE J Solid-State Circ 25(6):1379–1384
16. Yun Chiu I, Gray PR, Nikolic B (2004) A 14-b 12-MS/s CMOS pipeline ADC with over 100-dB SFDR. IEEE J Solid-State Circ 39(12):2139–2151
17. Bakker A, Huijsing JH (1996) Micropower CMOS temperature sensor with digital output. IEEE J Solid-State Circ 31(7):933–937
18. Razavi B (2001) Design of analog CMOS integrated circuits. McGraw-Hill, New York
19. Staszewski RB et al (2005) A digitally controlled oscillator in a 90 nm digital CMOS process for mobile phones. IEEE J Solid-State Circ 40(11):2203–2211
20. Gardner FM (1980) Charge-pump phase-locked loops. IEEE Trans Commun 28:1849–1858
21. Sun SY (1989) An analog PLL-based clock and data recovery circuit with high input jitter tolerance. IEEE J Solid-State Circ 24(2):325–330
22. Abidi AA, Meyer RG (1983) Noise in relaxation oscillators. IEEE J Solid-State Circ 18(6):794–802
23. Gierkink SLJ, van Tuijl Ed (AJM) (2002) A coupled sawtooth oscillator combining low jitter with high control linearity. IEEE J Solid State Circuits 37(6):702–710
24. Schoeff JA (1979) An inherently monotonic 12 bit DAC. IEEE J Solid-State Circ 14(6):904–911
25. van den Bosch A et al (2001) A 12 b 500 MSample/s current-steering CMOS D/A converter. In: IEEE ISSCC Dig. Tech. Papers, San Francisco, CA, pp 366–367

26. Zhang C, Makinwa KAA (2008) Interface electronics for a CMOS electrothermal frequency-locked-loop. IEEE J Solid-State Circ 43(7):1603–1608
27. Kashmiri SM, Makinwa KAA (2009) Measuring the thermal diffusivity of CMOS chips. In: Proceedings of the IEEE sensors, Christchurch, New Zealand, pp 45–48
28. Hamann HF et al (2007) Hotspot-limited microprocessors: direct temperature and power distribution measurements. IEEE J Solid-State Circ 42(1):56–65
29. Vermeersch B (2009) Thermal AC modelling, simulation and experimental analysis of microelectronic structures including nanoscale and high-speed effects. Ph.D. dissertation, University of Gent
30. Touloukian YS et al (1998) Thermophysical properties of matter, vol 10. Plenum, New York
31. McConnell AD, Goodson KE (2005) Thermal conduction in silicon micro- and nanostructures. Annu Rev Heat Trans 14:129–168
32. Ebrahimi J (1970) Thermal diffusivity measurement of small silicon chips. J Phys D Appl 3:236–239

Chapter 5
An Electrothermal Frequency Reference in Standard 0.7 μm CMOS

This chapter describes the design and implementation of an electrothermal (thermal-diffusivity-based) frequency reference in standard 0.7 μm CMOS. The reference locks the output frequency of a variable oscillator using a frequency-locked loop to the process-insensitive phase shift of an electrothermal filter. This is in turn a function of the thermal-diffusivity of silicon, which is temperature dependent. Therefore, the loop needs to be temperature-compensated. To do this, the digital output of an on-chip band-gap temperature sensor is applied to the digitally-assisted frequency-locked loop that was described in the previous chapter. The result is a frequency reference in a 0.7 μm standard CMOS whose output frequency is stable to within ±0.1% over the military temperature range ($-55°C$ to $125°C$).

5.1 Introduction

Contrary to all-silicon frequency references described in Chap. 2 [1–8], an electro-thermal (thermal-diffusivity-based) frequency reference embeds an electrothermal filter (ETF) into a frequency-locked loop (FLL). The design and implementation of a digitally-assisted FLL (DAFLL) was described in the previous chapter (Fig. 5.1). Feedback in the loop reassures that a digitally-controlled oscillator (DCO) oscillates at a frequency, f_{DCO}, where $\phi_{ETF} = \phi_{ref}$. As a result, f_{DCO} is determined solely by the ETF properties and is not affected by the DCO tolerances and temperature drift. In the DAFLL the output frequency of the DCO follows the $T^{-1.8}$ temperature dependence of D (Chap. 4). This means that to realize an output frequency which is stable over both process and temperature variations, some sort of temperature compensation is required.

The most straightforward method of temperature compensation would involve injecting a temperature-dependent signal into the loop. This can be done, by using an integrated temperature sensor that measures the instantaneous temperature of the die. With the digital nature of the DAFLL, it will be ideal if the temperature sensor

S.M. Kashmiri and K.A.A. Makinwa, *Electrothermal Frequency*
References in Standard CMOS, Analog Circuits and Signal Processing,
DOI 10.1007/978-1-4614-6473-0_5, © Springer Science+Business Media New York 2013

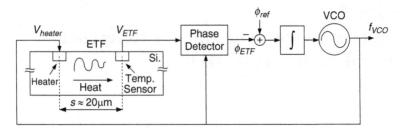

Fig. 5.1 Block-diagram of a simplified DAFLL

has a digital output, such that the temperature-compensating signal can be produced digitally. Such *smart temperature sensors* have been built based on the temperature dependence of bipolar transistors and called the band-gap temperature sensors. The inaccuracy of the temperature compensation scheme will add to the inherent inaccuracy of the reference, and so, designing a sufficiently accurate temperature sensor will be yet another challenge in the realization of an electrothermal frequency reference.

5.2   Temperature Compensation of Electrothermal Frequency-Locked Loops

As shown in Chap. 4, an electrothermal DAFLL can produce frequencies with device-to-device spreads less than ±0.25% [9–13]. However, this accuracy is defined with respect to a master curve with a $T^{-1.8}$ temperature dependence. This can be seen in Fig. 5.2, where the measured output frequency of the non-compensated DAFLL [13] (with fixed $\phi_{ref} = 90°$) is compared with a $T^{-1.8}$ fitted line (the graph is shown on a log-log scale). At room temperature this translates into a temperature coefficient of 0.3%/°C, which resembles the behavior of a temperature sensor more than that of a frequency reference.

Considering a simplified point-heater point-sensor model of an ETF, which is excited at a frequency f_{DCO} (see Fig. 5.1) and a die temperature T (in Kelvin), it was shown in Sect. 3.2 that:

$$\phi_{ETF} \propto \sqrt{f_{DCO}/D}. \tag{5.1}$$

with thermal-diffusivity D of silicon with temperature dependency [9, 10]:

$$D \propto 1/T^{1.8}. \tag{5.2}$$

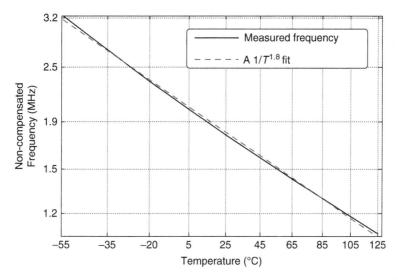

Fig. 5.2 Measured output frequency of the electrothermal FLL at a fixed $\phi_{ref} = 90°$ to a $T^{-1.8}$ fit on a log-log scale

As a result, for a fixed ϕ_{ref} of the DAFLL, the output frequency will be:

$$f_{DCO} \propto 1/T^{1.8}. \qquad (5.3)$$

This temperature dependency is illustrated in Fig. 5.3a, where the *measured* phase-frequency characteristics of an ETF ($s = 24$ μm) are plotted at various temperatures. It can be seen that for a constant $\phi_{ref} = 90°$ the DAFLL output frequency will be temperature-dependent. This is simply because the crossing point of the fixed horizontal ϕ_{ref} line with the characteristic of the ETF varies over the temperature. This intersecting point determines the locking frequency of the loop. It can be seen from Fig. 5.3a that the loop will have a temperature-dependent locking point, and therefore a temperature-dependent output frequency.

A suitable control knob, that can be used to keep the locking frequency of the loop constant, is already present in the DAFLL, and that is the phase reference input ϕ_{ref}. From the previous discussions it can be concluded that variations in the intersecting point of the loop's ϕ_{ref} with an ETF's characteristic influences the output frequency at any given temperature. This is because feedback in the loop will regulate the frequency such that $\phi_{ETF} = \phi_{ref}$.

The proposed temperature compensation method can be seen in Fig. 5.3b. Here, a constant value of f_{DCO} over temperature is achieved by ensuring that $\phi_{ref} \propto T^{0.9}$. This means that a temperature-dependent ϕ_{ref} keeps the intersection point at the same target frequency for all temperature values (in this case $f_{DCO} = 100$ kHz). Since this is a digital control of the loop, the required ϕ_{ref} can be produced by means of digital circuitry, which translates the temperature information into a $\phi_{ref} \propto T^{0.9}$ characteristic.

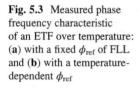

Fig. 5.3 Measured phase frequency characteristic of an ETF over temperature: (a) with a fixed ϕ_{ref} of FLL and (b) with a temperature-dependent ϕ_{ref}

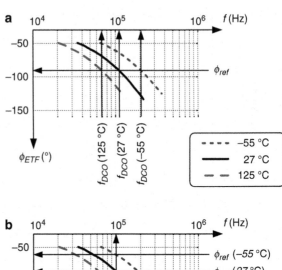

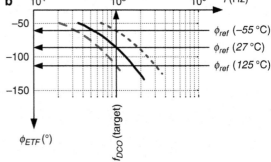

To give ϕ_{ref} the desired temperature dependence, a measure of the die temperature is required. This could be provided by an on-chip temperature sensor (TS) with a digital output. The digital output of the TS (Fig. 5.4) can then be translated digitally via a digital mapping scheme into a temperature-dependent $\phi_{ref}(T)$. The mapping function is unique for the whole batch and can be determined by batch-calibrating a number of devices.

The inaccuracy of the TS determines the accuracy of the temperature dependent $\phi_{ref}(T)$ and as a result contributes to the ultimate inaccuracy of the electrothermal frequency reference. Figure 5.5 shows the frequency error resulting from a fixed temperature measurement error. These results were obtained from a system-level simulation of the DAFLL shown in Fig. 4.3, using the thermal model [14] of the optimized ETF shown in Fig. 3.13 with $f_{drive} = 100$ kHz and over the temperature. These show that an absolute temperature measurement error of 0.1°C results in a frequency error of about 0.08%.

Intuitively, one might think of using a thermal-diffusivity-based temperature sensor (TD temperature sensor) [15] as the compensating TS of an electrothermal frequency reference. A TD temperature sensor measures the phase shift of an ETF driven at a constant frequency [see (5.1)]. However, it is exactly this constant frequency that must be generated! This issue resembles the so-called *chicken-and-egg* problem in the sense that to measure the temperature using a TD

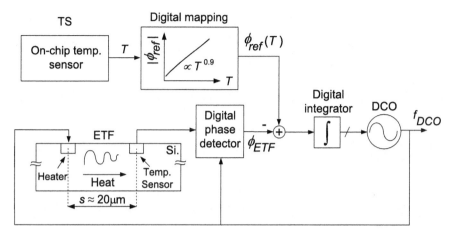

Fig. 5.4 Temperature compensation of DAFLL by means of an on-chip temperature sensor with digital output

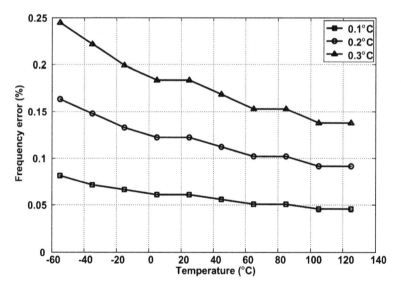

Fig. 5.5 Frequency error of the electrothermal frequency reference over temperature as a function of a fixed absolute temperature measurement error made by the TS (simulation)

temperature sensor, we need to have a well-known and stable frequency. Also, to make a well-known and stable frequency through an FLL, we need to know the temperature. This problem can be solved if the temperature sensor does not share (5.1) with the FLL, i.e. has a different temperature dependence. As a result, a different temperature sensing principle is required.

In the integrated circuit technology, various temperature-dependent elements can be used as temperature sensors, such as resistors, MOS transistors [16–18] or, the most commonly used, bipolar transistors (see [22], Chap. 3). Normally, these types of temperature sensors have an analog output signal in the form of a voltage or current. A survey of the TS literature [19–22] shows that *smart* band-gap temperature sensors with digital outputs and operating based on the temperature dependence of bipolar transistors are suitable candidates for the temperature compensation of electrothermal frequency references. On the one hand, their principle of operation is not fundamentally related to the FLL's frequency of operation. On the other hand, they are the most accurate class of temperature sensors, achieving accuracies in the order of 0.1°C (3σ) over wide temperature ranges (−55°C to 125°C) [19, 20]. Therefore, a temperature sensor (TS) architecture based on (see [22], Chap. 3), which utilizes the temperature dependence of substrate PNP transistors, was adopted.

5.3 Realization of an Electrothermal Frequency Reference in a 0.7 µm CMOS Process

The DAFLL of [13] was temperature-compensated with the help of a band-gap temperature sensor (TS). This was integrated in a test chip using standard 0.7 µm CMOS technology [23]. The TS was based on the chip described in [12] with a slight architectural modification to reduce its area and complexity. The resulting test chip produced an output frequency of 1.6 MHz with an inaccuracy of ±0.1% (= ±1,000 ppm) over the military temperature range, i.e. from −55°C to 125°C.

In this section, first the system-level considerations in the design of this frequency reference will be discussed. These concern the error contribution of the various building blocks to the ultimate inaccuracy of the reference. Furthermore, a detailed description of the TS design will be provided, which is followed by the experimental results from the test chip.

5.3.1 System-Level Design of the Reference

A simplified system-level diagram of the proposed electrothermal frequency reference is shown in Fig. 5.6. Its main components are a digitally-assisted electrothermal FLL (DAFLL) and a band-gap temperature sensor (TS). The DAFLL includes an ETF, a phase-domain $\Delta\Sigma$ modulator (PD$\Delta\Sigma$M), a 12-bit DCO, an up/down counter and a second-order digital $\Delta\Sigma$ modulator (D$\Delta\Sigma$M). The heater of the ETF is driven by a square wave at a frequency $f_{drive} = f_{DCO}/16$. The phase-shifted output of the ETF is then applied to the PD$\Delta\Sigma$M, which digitizes ϕ_{ETF} with respect to two phase shift references ϕ_0 and ϕ_1 [11]. The modulator bitstream output D_{out} is a digital representation of ϕ_{ETF}.

Band-gap Temperature Sensor (TS)

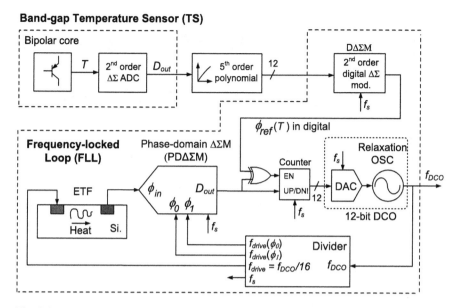

Fig. 5.6 Simplified system-level block diagram of the electrothermal frequency reference

The temperature dependent phase reference of the loop, $\phi_{ref}(T)$, is produced by the DΔΣM. Its bitstream output is subtracted from that of the PDΔΣM and the result is applied to a digital integrator, made of a 12-bit up/down counter. Subtraction of the two bitstreams is done through an XOR gate based on the method described in Chap. 4. The counter then drives the 12-bit input of the DCO. The DCO consists of a 12-bit current-steering DAC that tunes a relaxation oscillator [23]. The DAFLL is sampled at a frequency f_s (the PDΔΣM, the up/down counter, and the DCO), which is a sub-multiple of f_{drive}. The noise bandwidth of the system is thus determined by the length of the counter and the value of f_s. For a 12-bit counter and $f_s = f_{drive}/32$ the noise bandwidth of the loop is about 0.5 Hz.

The TS makes use of the temperature dependence of substrate PNP transistors. The temperature information extracted from a pair of PNPs is fed to a second-order ΔΣ ADC whose digital output is proportional to the die temperature. The temperature is then translated by means of a fifth-order polynomial into a 12-bit digital number that represents the $\phi_{ref}(T)$, resulting in a constant frequency of 1.6 MHz. This number is then converted by the DΔΣM into a noise-shaped single-bit bitstream. The quantization noise of both the PDΔΣM and DΔΣM are then suppressed by the digital integrator.

The ETF (described in Chap. 3) has a phase shift $\phi_{ETF} = 90°$ at $f_{drive} = 100$ kHz and room temperature. Its phase spread is lithography limited to about 0.1° in the target 0.7 μm CMOS process [11]. In the electrothermal frequency reference, this phase spread translates into an output frequency error of about ±0.25%. Therefore, the other blocks in the system must be designed to contribute significantly less error. The major sources of error are then the phase error introduced by the PDΔΣM [11] and the temperature sensing inaccuracy of the TS. The simulation results shown in Fig. 5.5 imply that temperature sensing inaccuracy of 0.1°C is required.

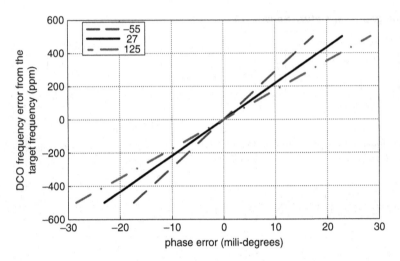

Fig. 5.7 The frequency error due to phase error introduced by the PDΔΣM

Table 5.1 The error contribution of various building blocks to the output frequency of the electrothermal frequency reference

Building block	Inaccuracy contribution to output frequency	Tolerated inaccuracy	Required range	Required resolution
PDΔΣM	15 m ° phase error → 500 ppm	10 m °	ETF phase over the military temperature range (65 ° → 105 °)	>12-bits
TS	0.1°C → 800 ppm	0.1°C	From −55°C to 125°C	<50 m°C
DCO	–	–	Enough to cover its ±40% spread from nominal frequency of 1.6 MHz over PVT	0.05% (800 Hz)

The mechanisms leading to phase measurement error in the PDΔΣM are discussed extensively in Chap. 4. To investigate their overall effect on the output frequency of the electrothermal frequency reference, system-level simulations were performed on the block diagram of Fig. 5.6 using a model for the optimized ETF shown in Fig. 3.13 with $f_{drive} = 100$ kHz and different levels of phase error introduced by the PDΔΣM at various temperatures. Figure 5.7 shows that for a worst-case frequency error of 0.05% at −55°C, a maximum phase error of 15 milli-degrees can be tolerated.

The system level simulations studying the effects of the TS and the PDΔΣM on the output frequency inaccuracy lead to a set of specifications required from the various system blocks. The specifications required to achieve a target inaccuracy of 0.1% are summarized in Table 5.1. As discussed in Chap. 4, the DAFLL has already been designed for the error budget described in the first and third rows of Table 5.1. Therefore, to build a complete electrothermal (thermal-diffusivity-based) frequency reference, the band-gap TS is the only remaining block to be considered. The following sub-sections describe the design of the TS.

5.3.2   The Band-Gap Temperature Sensor Design

5.3.2.1   System-Level Design

The output of the TS is the ratio of a temperature-dependent parameter and a temperature-independent parameter. The former is a proportional-to-absolute temperature voltage (PTAT), while the latter is the well-known band-gap voltage [24]. This voltage can be made by combining a PTAT voltage with a complementary-to-absolute temperature voltage (CTAT). A band-gap reference, therefore, has both of the terms a temperature sensor requires: a temperature-independent voltage and a PTAT voltage. To produce these voltages, bipolar transistors are normally used. The base emitter voltage of a bipolar transistor V_{BE} is a function of its collector current I_C and its saturation current I_S [19, 25]:

$$V_{BE} = \frac{kT}{q} \ln\left(\frac{I_C}{I_S}\right). \tag{5.4}$$

where k is Boltzmann's constant, T is the absolute temperature, and q is the electron charge. The value of V_{BE}, which has a CTAT dependence with a sensitivity of about -2 mV/°C, can be extrapolated to the silicon band-gap voltage of 1.2 V at zero Kelvin. To produce a PTAT voltage, two identical bipolar transistors can be biased at different current densities with collector currents I_{C1} and I_{C2} related as:

$$I_{C2} = p \cdot I_{C1}. \tag{5.5}$$

Through (5.4) we can expand the difference between the base emitter voltages V_{BE1} and V_{BE2} of these transistors to:

$$V_{BE2} - V_{BE1} = \frac{kT}{q} \ln\left(\frac{p \cdot I_{C1}}{I_S}\right) - \frac{kT}{q} \ln\left(\frac{I_{C1}}{I_S}\right) = \frac{kT}{q} \ln(p). \tag{5.6}$$

which shows that the difference between the base emitter voltages of these transistors is proportional to the absolute temperature. This PTAT voltage is thus an accurate measure of the temperature, such that its accuracy is only determined by the accuracy of the current ratio of the bipolar transistors' bias currents [19].

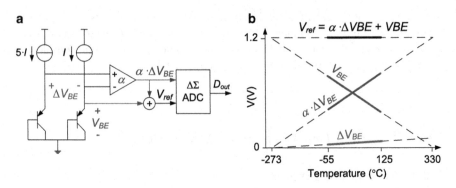

Fig. 5.8 (a) System-level block diagram of the TS and (b) the PTAT and band-gap reference voltages over the temperature

In this work, two diode-connected substrate PNP transistors are biased at a current ratio of 5:1 (see Fig. 5.8a). This causes the difference between their base emitter voltages ΔV_{BE} to be:

$$\Delta V_{BE}(T) = \frac{kT}{q} \ln(5). \tag{5.7}$$

If an appropriately scaled PTAT voltage $\alpha \cdot \Delta V_{BE}$ is combined with a V_{BE}, a band-gap reference voltage, V_{REF}, can be made (see Fig. 5.8b). In this design, a gain factor $\alpha = 16$ is used. The PTAT voltage and the band-gap voltage are then both fed into a $\Delta\Sigma$ ADC (Fig. 5.8a) consisting of a $\Delta\Sigma$ modulator and a decimation filter [26].

The $\Delta\Sigma$ modulator consists of a charge-balancing loop-filter and a clocked quantizer (Fig. 5.9). At every clock cycle, the bitstream polarity determines whether $\alpha \cdot \Delta V_{BE}$ ($bs = 0$), or $-V_{BE}$ ($bs = 1$) should be input to the loop-filter (Fig. 5.9). This happens by means of charge packets (voltage-to-charge transfer and vice versa are not shown in the figure). Due to feedback, the average input to the loop filter should be zero and so the charge added by $\alpha \cdot \Delta V_{BE}$ will be balanced by that removed by $-V_{BE}$. The bitstream average denoted by μ can be expressed as:

$$(1 - \mu) \cdot \alpha \cdot \Delta V_{BE} = \mu \cdot V_{BE}. \tag{5.8}$$

Solving this equation gives:

$$\mu = \frac{\alpha \cdot \Delta V_{BE}}{\alpha \cdot \Delta V_{BE} + V_{BE}} = \frac{V_{PTAT}}{V_{REF}}. \tag{5.9}$$

which is a digital representation of the die temperature. The value of μ ranges between about 0.4 and 0.7 over the military temperature range [19].

Fig. 5.9 Charge-balancing
scheme in the ΔΣ modulator
of the TS

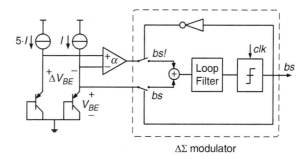

ΔΣ modulator

Fig. 5.10 The simplified
block-diagram of the feed-
forward second-order ΔΣ
modulator

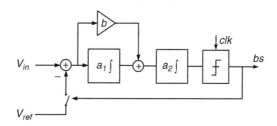

The level of inaccuracy required from the TS means that all its error sources
have to be reduced such that their contribution to the total inaccuracy is well below
0.1°C. This includes the offset in the readout circuitry of ΔV_{BE}, mismatch in the
current ratio of 1:5, and error in the gain factor α [19]. If the contribution of these
error sources is negligible, the only remaining source of inaccuracy in the TS will
be the process spread of V_{BE}, which, as in [19], is corrected by a single PTAT trim.
This is done by varying the bias current of the bipolar device that produces V_{BE}.
This trimming procedure is discussed further in the next section.

The ΔΣ modulator of the TS incorporates a second-order loop filter (see
Fig. 5.10) based on the same feed-forward topology used in [12]. The loop filter
includes two integrators with gains a_1 and a_2 and a feed-forward coefficient b. The
feed-forward path helps reduce the signal swing at the first integrator output of the
loop filter [27]. The integrator gains a_1 and a_2, as well as the feed-forward coefficient b, were chosen to be 1/4, 1/2, and 1/2, respectively [19].

The charge-balancing scheme shown in Fig. 5.9 can be combined with the feed-
forward second order loop's block-diagram shown in Fig. 5.10. This will result in
the block diagrams shown in Fig. 5.11, illustrating the ΔΣ modulator during the two
charge balancing phases. When bs = 0, the value of ΔV_{BE} is applied to the
modulator, and the gain factor α is implemented by the first integrator's gain as
well as the feed-forward coefficient, such that they become $\alpha \cdot a_1$ and $\alpha \cdot b$,
respectively. During the V_{BE} cycle when bs = 1, the coefficients are switched
back to a_1 and b.

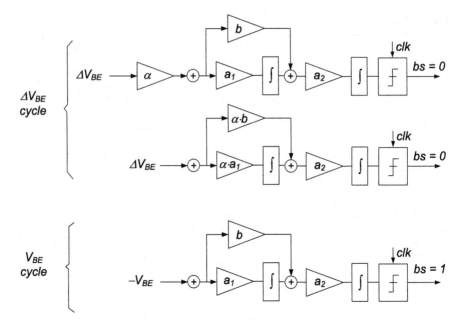

Fig. 5.11 The charge balancing phases in the system-level block diagram of the second-order feed-forward ΔΣ modulator

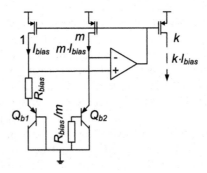

Fig. 5.12 Simplified PTAT bias current generator

5.3.2.2   Circuit Design

The TS employs a bias circuit that provides a bipolar core with a PTAT bias current [(see [22], Chap. 3), 12]. The bias circuit (Fig. 5.12) has been directly adopted from [12] and so it will be briefly discussed here. The two PNP transistors Q_{b1} and Q_{b2} are biased at a 1:m ($m = 10$) current ratio and hence their ΔV_{BE} is a PTAT voltage [see (5.7)]. The feedback loop enforces virtual ground at the inputs of the opamp.

Considering the forward β of the transistors to be β_F and considering the effect of their base currents [19]:

$$I_{bias} = \frac{\beta_F + 1}{\beta_F} \frac{\Delta V_{BE}}{R_{bias}}. \tag{5.10}$$

considering that the substrate PNPs in the bipolar core are biased through their emitters (see Fig. 5.8), their V_{BE} will be a function of their base current and hence β_F dependent. The advantage of biasing such transistor with a bias current defined by (5.10) is that the β_F dependence cancels (assuming matching between the PNPs in the bias circuit and the transistors in the bipolar core) [19].

The bipolar core generates the voltages ΔV_{BE} and V_{BE}. A charge-balancing second-order switched-capacitor $\Delta\Sigma$ modulator produces the ratiometric measurement value μ derived in (5.9). The charge-balancing phases of this modulator are described below. Figure 5.13a illustrates a simplified circuit diagram of the bipolar core as well as the switched-capacitor second-order loop filter of the modulator, when $bs = 0$. In this case, a set of six PMOS current sources provides copies of the current produced in the bias circuit, each with a nominal value of 1 μA. A set of switches is used to direct five of the six currents to one of the two bipolar transistors (Q_L or Q_R) while the remaining current goes to the other transistor, producing the 5:1 current ratio. Mismatch between the current sources leads to an error in this ratio, which is eliminated by means of dynamic element matching (DEM). i.e. by periodically alternating the current source that generates the unit current, so that the mismatch errors are averaged out during the course of a conversion [19].

As in [19] the correlation between the cyclic behavior of the DEM and the limit cycles of the $\Delta\Sigma$ modulator had to be broken. Otherwise, this leads to the fold-back of quantization noise and loss of performance in the modulator. To do so, the DEM scheme is only updated during the ΔV_{BE} phase and its state is frozen when the modulator has to sample the V_{BE}. The timing of this bitstream-controlled DEM algorithm can be seen in Fig. 5.14a.

The multiplexer MUX feeds ΔV_{BE}, as the voltage $V_{\Delta\Sigma}$, to the sampling capacitors C_S (each 5 pF) of the loop filter's first integrator (Fig. 5.13a). This voltage is sampled during phase ϕ_1 and integrated during phase ϕ_2 (Fig. 5.14a) on the integrating capacitors C_{int1} (each 20 pF). In contrast to the approach used in [12, 19], the gain factor $\alpha = 16$ is implemented here by sampling and then integrating ΔV_{BE} 16 times with a single sampling capacitor (see Fig. 5.14a). As a result, the total integrated charge in this phase is given by:

$$Q_{(bs=0)} = 16 \cdot C_s \cdot \left(\Delta V_{BE,RL} + \Delta V_{BE,LR}\right). \tag{5.11}$$

where $\Delta V_{BE,RL} = V_{BE,R} - V_{BE,L}$ and $\Delta V_{BE,LR} = V_{BE,L} - V_{BE,R}$ (see Fig. 5.13a). Since the gain factor α, also needs to be applied to the feed-forward coefficient b during the ΔV_{BE} cycle, the feed-forward capacitor C_{FF} (each 1 pF) is also switched 16 times with the same timing the sampling capacitors are switched during the

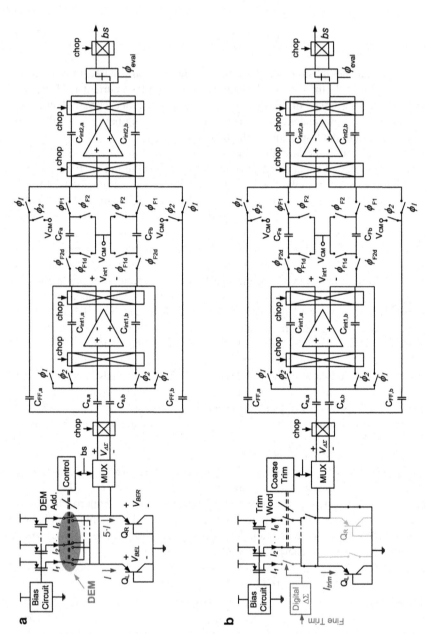

Fig. 5.13 Switched capacitor loop-filter of the $\Delta\Sigma$ modulator and the bipolar core, sampling ΔV_{BE} when bs = 0 (**a**) and sampling V_{BE} when bs = 1 (**b**)

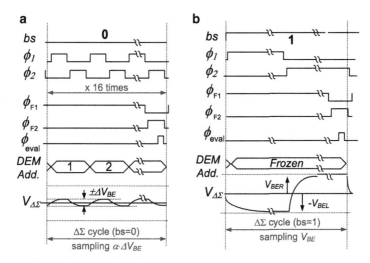

Fig. 5.14 Corresponding timing diagrams during a $\Delta\Sigma$ cycle while sampling ΔV_{BE} when bs $= 0$ (**a**) and sampling V_{BE} when bs $= 1$ (**b**)

phases ϕ_1 and ϕ_2 (see Fig. 5.13a). At the end of every ΔV_{BE} cycle, the integrated charge in the first integrator is transferred to the capacitor C_F (each 2 pF) forming the sampling capacitor of the second integrator. This is discharged to the second integrator's integrating cap C_{int2} (each 4 pF). The modulator's coefficients can be calculated as:

$$a_1 = \frac{C_S}{C_{int1}} = \frac{5pF}{20pF} = \frac{1}{4}, \ a_2 = \frac{C_F}{C_{int2}} = \frac{2pF}{4pF} = \frac{1}{2}, b = \frac{C_{FF}}{C_F} = \frac{1pF}{2pF} = \frac{1}{2}. \quad (5.12)$$

Figure 5.13b illustrates a simplified circuit diagram of the bipolar core and the loop filter, when $bs = 1$. In this phase, a variable bias current I_{trim} is applied to either of the bipolar transistors Q_L or Q_R, producing a trimmable V_{BE} voltage. Of the six PMOS current sources, five are controlled by a coarse-trim word to form the coarse part of I_{trim}, which can be set from 1 to 5 μA. The sixth current source is switched on and off to fine-tune the current, using the bitstream output of a digital $\Delta\Sigma$ modulator with a fine-trim word as the input. The switching of the current source is controlled by a first order digital $\Delta\Sigma$ modulator with a full scale between 0 and 255. Like (see [22], Chap. 3) the digital $\Delta\Sigma$ modulator is clocked only when the main second order $\Delta\Sigma$ modulator is in the V_{BE} mode and otherwise its accumulator is frozen during the ΔV_{BE} phase. This will prevent the tonal behavior of this modulator from interacting with the quantization noise of the main modulator and causing intermodulation products, which will result in the fold-back of quantization noise [28].

The multiplexer MUX then feeds either the $V_{BE,L}$ or $V_{BE,R}$ to the modulator's sampling capacitors (see Fig. 5.14b). The total integrated charge in this phase is:

$$Q_{(bs=1)} = C_s \cdot \left(V_{BE,R} + V_{BE,L}\right). \quad (5.13)$$

The charge balancing requires (5.11) to be equal to (5.13), which results in the equivalent of the charge-balancing equation (5.8):

$$(1 - \mu) \cdot 16 \cdot C_s \cdot \left(\Delta V_{BE,RL} + \Delta V_{BE,LR}\right) = \mu \cdot C_s \cdot \left(V_{BE,R} + V_{BE,L}\right). \tag{5.14}$$

where μ is the bitstream average. This shows that the effective value of the gain factor α applied to ΔV_{BE} is 16, which is fixed by the number of times ΔV_{BE} is sampled. The structure of the first and second integrators' opamps as well as the quantizer of the $\Delta\Sigma$ modulator remain exactly the same as (see [22], Chap. 3), and thus are not further discussed here.

5.3.2.3   Implications of Single Capacitor Sampling

As discussed in the earlier sections, the gain factor α was implemented in [19] by using multiple capacitors to sample ΔV_{BE}, while using only a single capacitor to sample V_{BE}. As a result, capacitor mismatch limited the accuracy of α and DEM was required. In this work, the gain factor α was implemented by sampling ΔV_{BE} multiple times on the same capacitor used to sample V_{BE}, thus guaranteeing the accuracy of α. The associated reduction in the number of sampling capacitors, as well as the elimination of the digital circuitry needed for DEM, leads to both a reduction of 30% in the area of the TS compared to that of [19], as well as a significant reduction in its complexity.

These improvements, however, come at the cost of increased charge-injection-related offsets, since compared to [19], the number of switching actions per ΔV_{BE} cycle of the $\Delta\Sigma$ modulator is increased by a factor 8. However, this is eliminated by the system-level chopping scheme described in [19]. To ensure the complete elimination of the chopping tone, two full periods of the system-level chopping clock (see the *chop* signal in Fig. 5.13) are applied during every conversion of the TS, such that it will be completely eliminated by the sinc^2 decimation filter following the modulator [19].

The input-referred noise of the single capacitor sampling scheme needs to be investigated. This is because it directly determines the resolution of the TS. It is also important to investigate the settling behavior of the single capacitor sampling approach, as this will impact the accuracy of the TS.

First we compare the kT/C noise performance of the approach in [19] with the single capacitor solution proposed here. In the case of [19, (see [22], Chap. 3)], a gain factor α of 16 was implemented by sampling the ΔV_{BE} produced by the bipolar core two times on eight sampling capacitors each with a value of 8·C_S. The resultant charge corresponding to the sampled kT/C noise can be determined as:

$$V_n = \sqrt{2 \cdot \frac{kT}{(8C_S)}} \Rightarrow q_n = (8C_S)\sqrt{2 \cdot \frac{kT}{(8C_S)}} = \sqrt{16 \cdot kTC_S}. \tag{5.15}$$

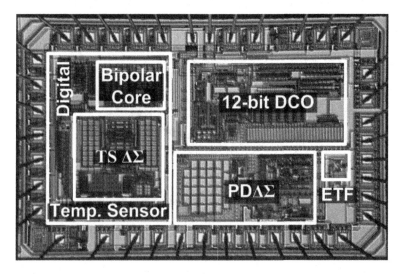

Fig. 5.15 Chip photomicrograph of the electrothermal frequency reference in 0.7 μm CMOS

while in the case of a single capacitor with a value of C_S sampling ΔV_{BE} for 16 times, the corresponding noise charge is:

$$V_n = \sqrt{16 \cdot \frac{kT}{C_S}} \Rightarrow q_n = C_S \sqrt{16 \cdot \frac{kT}{C_S}} = \sqrt{16 \cdot kTC_S}. \qquad (5.16)$$

This shows that both approaches have exactly the same amount of noise charge at the end of a ΔV_{BE} cycle. Since in both cases this noise charge is compared to the signal charge due to the sampling of ΔV_{BE}, which is $16 \cdot C_S \cdot \Delta V_{BE}$ for both cases, the signal-to-noise ratio, and therefore the resolution of the TS, remain the same.

Incomplete settling of the voltage across the sampling capacitors leads to an error in the gain α [19]. For a given settling error a certain number of time constants τ, determined by the PTAT bias current of the bipolar transistor and the sampling capacitor value will be required [19]. Comparing the multiple capacitor and the single capacitor approaches for a fixed $\Delta \Sigma$ cycle length, implies that the single capacitor approach requires eight times faster settling. However, the capacitance value is also eight times smaller and therefore the required settling time is, in principle, proportionally smaller.

5.3.3 Experimental Results

The electrothermal (thermal-diffusivity-based) frequency reference including the DAFLL and the TS was implemented in a standard 0.7 μm CMOS process (Fig. 5.15) [23]. The chip has a die size of 6.75 mm², and consumes 7.8 mW

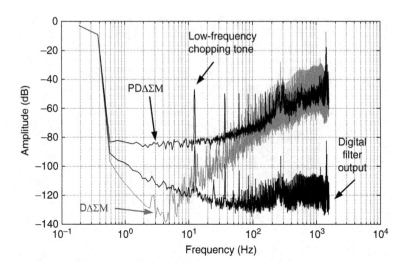

Fig. 5.16 The measured output spectrum of the PDΔΣM (the low-frequency chopping tone at 12 Hz), the second-order DΔΣM, and the FLL digital filter (16384-point FFT, Hanning window, 8× averaged)

from a 5 V supply, 2.5 mW of which is dissipated in the ETF. The DAFLL is exactly the same as the one described in Chap. 4 [13], and the band-gap TS, was added to the left side of the DAFLL in layout. For flexibility, the up/down counter, frequency dividers, the DΔΣM, the digital ΔΣ modulator producing the fine trimming word for the TS, and the decimation filter of the TS were realized off-chip in an FPGA. For characterization, 16 devices were packaged in ceramic DIL packages and placed in a climate chamber, in good thermal contact with a platinum PT-100 thermistor, which was calibrated to 20 mK.

The output spectrum of the PDΔΣM is shown in Fig. 5.16, as well as the spectrum of the signal at the input of DCO. It confirms that the digital integrator effectively suppresses the low frequency tone caused by chopping, as well as the high frequency quantization and truncation noise due to the PDΔΣM and DΔΣM.

The compensation polynomial that maps the die temperature onto a temperature-dependent phase reference $\phi_{ref}(T)$ for the FLL was extracted by *batch-calibrating* 16 devices over temperature. First, the FLL was calibrated by determining the values of $\phi_{ref}(T)$ (Fig. 5.17a) that correspond to the target frequency of 1.6 MHz over temperature. During this process, the digital output of the TS (μ as a function of temperature), was measured at a fixed coarse trim current of $I_{trim} = 2$ μA (Fig. 5.17b). The device-to-device spread of ϕ_{ref} with respect to the FLL's master curve (Fig. 5.17a), i.e. the FLL's untrimmed phase error, is ±0.1° (Fig. 5.18a). The measured device-to-device spread in μ with respect to the TS master curve (Fig. 5.17b) corresponds to ±0.2°C (Fig. 5.18b), which is a measure of its untrimmed inaccuracy. Based on these master curves, a fifth-order polynomial was derived that maps the output of the TS to the 12-bit digital input to DΔΣM, in order to produce ϕ_{ref}. This polynomial (Fig. 5.19) is common to all devices.

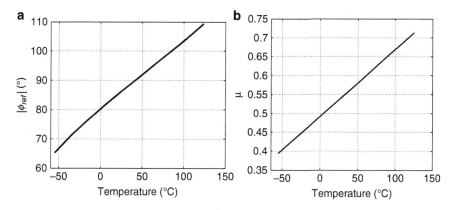

Fig. 5.17 (a) The measured ϕ_{ref} that tunes f_{DCO} to 1.6 MHz as a function of temperature and (b) μ the measured output of TS varying between 0 and 1 over temperature

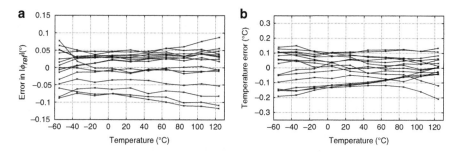

Fig. 5.18 Measurements of (a) the untrimmed error of ϕ_{ref} over temperature and (b) the untrimmed inaccuracy of the TS

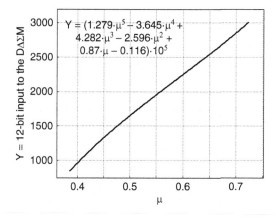

Fig. 5.19 The fifth-order polynomial that maps the output of TS (μ) to the 12-bit digital input of the DΔΣM

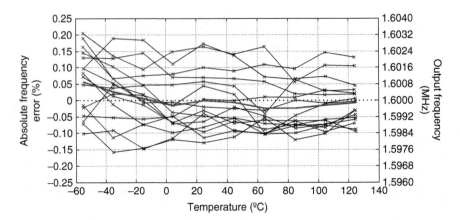

Fig. 5.20 The measured absolute frequency error of 16 devices from the target frequency of 1.6 MHz over the military range before trimming (the *bold line* indicates the average error)

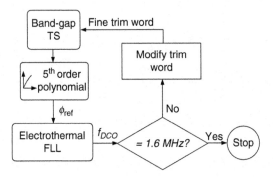

Fig. 5.21 Trimming algorithm of the electrothermal frequency reference

After the calibration procedure, temperature compensation was applied to the FLL by running the temperature sensor at a conversion rate of 2 Hz, in line with the FLL's 0.5 Hz bandwidth. The measured output frequency of 16 devices over temperature, without applying any individual trimming, shows an absolute frequency error of ±0.2% from the target frequency of 1.6 MHz (Fig. 5.20).

As in [19], the spread in the PNP's saturation current can be compensated by a single PTAT trim. Since the ETF's spread is also PTAT in nature (Fig. 5.18a), both sources of error can be compensated by the same trimming knob. This is the fine trim of the bias current of the PNP devices, I_{trim}. The trimming procedure follows the algorithm shown in Fig. 5.21. For a given device, a fine-trim word is applied to the TS whose output produces then a ϕ_{ref} for the FLL via the fifth-order polynomial. The fine-trim word is adjusted until the resulting output frequency is equal to the target frequency of 1.6 MHz. The corresponding trim word is then stored and used over the entire temperature range. Since the worst-case temperature coefficient for

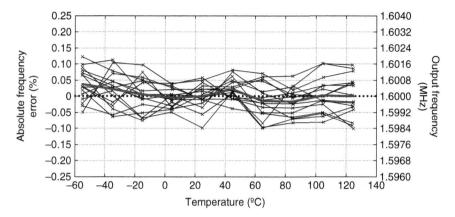

Fig. 5.22 Measured absolute frequency error of 16 devices from the target frequency of 1.6 MHz over the military range after a single room temperature trimming (the *bold line* is the average error)

Table 5.2 Summary of the power dissipation, size and the measured performance of the main building blocks of the electrothermal frequency reference

Building block	Power dissipation	Percent of the total power	Size (mm^2)	Major performance metrics measured
PDΔΣM	2.5 mW	32%	0.845	Combined untrimmed phase
ETF	2.5 mW	32%	0.022	inaccuracy: ±0.1 °
DCO	2.5 mW	32%	1.125	Tuning range: 350 kHz ~ 4 MHz
TS	0.3 mW	4%	1.3	Untrimmed inaccuracy: ±0.2°C
Complete chip	7.8 mW	100%	6.75	Frequency inaccuracy after a single room-temperature trim: ±0.1%

the reference is only ±11.2 ppm/°C, its temperature does not need to be stable during this trimming procedure, greatly simplifying the trimming process. After a single trim at room temperature the electrothermal frequency reference achieves an absolute output frequency inaccuracy of ±0.1% with σ = ±0.05% (Fig. 5.22). Table 5.2 summarizes the performance of the various building blocks of the system. Table 5.3 compares the performance of the electrothermal frequency reference with other state-of-the-art temperature-compensated oscillators [2, 5, 7].

5.4 Conclusions

The realization and characterization of a fully-integrated electrothermal (thermal-diffusivity-based) frequency reference, implemented in a standard 0.7 μm CMOS process was discussed in this chapter. A digitally-assisted electrothermal

Table 5.3 Performance comparison of the electrothermal frequency reference with state-of-the-art temperature-compensated oscillators

Reference	This work	[5] and [6]-LC	[7]-ring	[2]-RC
Frequency	1.6 MHz	24 MHz	7.03 MHz	6 MHz
Supply voltage (V)	5	1.8	2.5	1.2
Power consumption	7.8 mW	<4 mW	1.5 mW	66 μW
Technology	0.7 μm	0.13 μm	0.25 μm	65 nm
Die size (mm^2)	6.75	0.8	1.6	0.03
Temp. range (°C)	−55 ~ 125	0 ~ 70	−40 ~ 125	0 ~ 120
Inaccuracy	±0.1%	±0.005% to ±0.03%	±1.84%	±0.9%
Number of samples	16	Product	94	6
Period jitter (rms)	312 ps	<1 ps	NA	NA
Temp. coeff. (ppm/ °C)	±11.2	±1.4 to ±8.6	±50.9	86.1

NA not available

frequency-locked loop (DAFLL) locks the output frequency of a digitally-controlled oscillator (DCO) to the process-insensitive phase shift of an electrothermal filter (ETF). The ETF's phase shift is a function of its geometry and the thermal-diffusivity constant D of bulk silicon. The temperature dependence of D leads to a temperature dependent frequency, which in this work was compensated for by measuring the temperature of the die with an on-chip band-gap temperature sensor and injecting the temperature information to the DAFLL digitally.

The reference has an output frequency of 1.6 MHz, dissipates 7.8 mW from a 5 V supply and achieves an absolute inaccuracy of ±0.1% over the military temperature range (−55°C to 125°C) with a single room-temperature trim. The worst-case temperature coefficient of ±11.2 ppm/°C allows for trimming without temperature stabilization, which simplifies the trimming procedure and thus reduces trimming costs. Since the DCO has been designed with sufficient tuning range and resolution, the reference's ultimate inaccuracy is determined by the inaccuracy of the temperature sensor used for temperature compensation, and by the inaccuracy of the ETF's phase shift, which is, to first order, determined by the accuracy of the lithography in a 0.7 μm CMOS process.

References

1. Tokunaga Y et al (2010) An on-chip CMOS relaxation oscillator with voltage averaging feedback. IEEE J Solid-State Circ 45(6):1150–1158
2. De Smedt V et al (2009) A 66 μW 86 ppm/ °C fully-integrated 6 MHz Wienbridge oscillator with a 172 dB phase noise FOM. IEEE J Solid-State Circ 44(7):1990–2001
3. McCorquodale MS et al (2007) A monolithic and self-referenced RF LC clock generator compliant with USB 2.0. IEEE J Solid-State Circ 42(2):385–399
4. McCorquodale MS et al (2008) A 0.5-to-480 MHz self-referenced CMOS clock generator with 90 ppm total frequency error and spread-spectrum capability. In: IEEE ISSCC Dig. Tech. Papers, San Francisco, CA, pp 350–351

5. McCorquodale MS et al (2010) A silicon die as a frequency source. In: Proceedings of the IEEE international frequency control symposium, Newport Beach, California, June 2010, pp 103–108

6. McCorquodale MS et al (2011) A history of the development of CMOS oscillators: the dark horse in frequency control. In: IEEE international frequency control symposium, San Francisco, CA, pp 437–442

7. Sundaresan K et al (2006) Process and temperature compensation in a 7-MHz CMOS clock oscillator. IEEE J Solid-State Circ 41(2):433–442

8. Foley DJ et al (2006) CMOS DLL-based 2-V 3.2-ps jitter 1-GHz clock synthesizer and temperature-compensated tunable oscillator. IEEE J Solid-State Circ 36:417–423

9. Makinwa KAA, Snoeij MF (2006) A CMOS temperature-to-frequency converter with an inaccuracy of less than ±0.5 °C (3σ) from −40 °C to 105 °C. IEEE J Solid-State Circ 41(12):2992–2997

10. Szekely V (1994) Thermal monitoring of microelectronic structures. Microelectron J 25(3):157–170

11. Kashmiri SM et al (2009) A temperature-to-digital converter based on an optimized electro-thermal filter. IEEE J Solid-State Circ 44(7):2026–2035

12. Pertijs MAP et al (2005) A CMOS smart temperature sensor with a 3σ inaccuracy of ±0.1 °C from −55 °C to 125 °C. IEEE J Solid-State Circ 40(12):2805–2815

13. Kashmiri SM, Makinwa KAA (2009) A digitally-assisted electrothermal frequency-locked loop. In: Proceedings of the 35th ESSCIRC, Athens, Greece, pp 296–299

14. Kashmiri SM, Makinwa KAA (2009) Measuring the thermal diffusivity of CMOS chips. In: Proceedings of the IEEE sensors, Christ church, New Zealand, pp 45–48

15. van Vroonhoven CPL et al (2010) A thermal-diffusivity-based temperature sensor with an untrimmed inaccuracy of ±0.2 °C (3σ) from −55 °C to 125 °C. In: IEEE ISSCC Dig. Tech. Papers, San Francisco, CA, pp 314–315

16. Chen P et al (2009) A time-domain sub-micro watt temperature sensor with digital set-point programming. IEEE Sens J 9(12):1639–1646

17. Tuthill M (1998) A switched-current, switched-capacitor temperature sensor in 0.6 μ CMOS. IEEE J Solid-State Circ 33(7):1117–1122

18. Meijer GCM et al (2001) Temperature sensors and voltage references implemented in CMOS technology. IEEE Sens J 1(3):225–234

19. Pertijs MAP, Huijsing JH (2006) Precision temperature sensors in CMOS technology. Springer, Dordrecht

20. Aita AL et al (2009) A CMOS smart temperature sensor with a batch-calibrated inaccuracy of ±0.25 °C (3σ) from −70 °C to 130 °C. In: IEEE ISSCC Dig. Tech. Papers, San Francisco, CA, pp 342–343

21. Sebastiano F (2010) A 1.2-V 10-μW NPN-based temperature sensor in 65-nm CMOS with an inaccuracy of 0.2 °C (3σ) from −70 C to 125 °C. IEEE J Solid-State Circ 45(12):2591–2601

22. Souri K et al (2010) A CMOS temperature sensor with an energy-efficient zoom ADC and an inaccuracy of ±0.25 °C (3σ) from −40 °C to 125 °C. In: ISSCC Dig. Tech. Papers, February 2010, San Francisco, CA, pp 310–311

23. Kashmiri SM et al (2010) A thermal-diffusivity-based frequency reference in standard CMOS with an absolute inaccuracy of ±0.1 % from −55 °C to 125 °C. IEEE J Solid-State Circ 45(12):2510–2520

24. Brokaw AP (1974) A simple three-terminal IC bandgap reference. IEEE J Solid-State Circ 9(6):388–393

25. Gray PR, Meyer RG (2001) Analysis and design of analog integrated circuits, 4th edn. Wiley. New York, ISBN 0-471-32168-0

26. Schreier R, Temes GC (2005) Understanding delta-sigma data converters. Wiley, Chichester

27. Temes G, Steensgaard J (2007) Structural optimization and scaling of delta-sigma modulators. In: Lecture notes of MEAD advanced engineering course on delta sigma modulators, Lausanne, Switzerland

28. Pertijs MAP et al (2004) Bitstream trimming of a smart temperature sensor. In: Proceedings of the IEEE sensors, Vienna, Austria, pp 904–907

Chapter 6
A Scaled Electrothermal Frequency Reference in Standard 0.16 μm CMOS

The previous chapter described an implementation of an electrothermal (thermal-diffusivity-based) frequency reference. A prototype in a standard 0.7 μm CMOS technology demonstrated the feasibility of such references. The inaccuracy of its 1.6 MHz output frequency was ±0.1% over the military temperature range, and its cycle-to-cycle jitter was about 400 ps (rms). This chapter describes the implementation of a scaled electrothermal frequency reference, whose performance is improved by means of scaling.

The performance of an electrothermal frequency reference is mainly determined by its electrothermal filter (ETF), which forms the heart of the reference and whose characteristics are mainly determined by its geometry. In particular, the accuracy of the filter's phase response will be determined by lithographic inaccuracy. This will then determine the accuracy of the reference's output frequency. As CMOS processes scale, lithographic accuracy improves, thus improving ETF accuracy. To investigate this concept, a prototype has been designed in a more advanced 0.16 μm CMOS process. Compared to the previous prototype, the scaled reference achieves 10× higher frequency, 7× less jitter, 3.7× less power, and 12× less chip area, while maintaining the same level of accuracy.

6.1 Introduction

CMOS scaling has been mainly driven by the increasing demand for higher performance microprocessors. Scaling allows for the inclusion of more transistors on a single die, which results in more functionality. Furthermore, scaling results in shorter transistor lengths and thinner gate oxides, which both result in faster devices [1, 2]. Also, the reduction of supply voltages helps reducing the power consumption of digital circuitry. Unfortunately, these modifications limit the performance of analog circuits. On the one hand, the reduced supply voltage limits the signal swings, while, on the other hand, the short channel effects reduce the intrinsic gain of transistors [3].

S.M. Kashmiri and K.A.A. Makinwa, *Electrothermal Frequency*   153
References in Standard CMOS, Analog Circuits and Signal Processing,
DOI 10.1007/978-1-4614-6473-0_6, © Springer Science+Business Media New York 2013

Although CMOS scaling makes the design of the analog circuits of an electrothermal frequency reference more challenging; however, it can significantly improve the performance of its electrothermal filter [4]. In scaled CMOS, the lithography should resolve smaller feature sizes, which requires more accuracy. Furthermore, the reduced feature sizes allow for implementation of smaller ETF structures, i.e. smaller heater-thermopile distances and smaller thermopile dimensions. These help increase the output signal of an ETF and reduce its thermal noise, which eventually improve the jitter performance of the frequency reference (see Chap. 3). Furthermore, smaller dimensions reduce the filter's thermal phase, which enables the use of higher excitation frequencies.

To implement the scaled electrothermal frequency reference, a single-poly, 5-metal baseline 0.16 μm CMOS process with 1.8 V supply has been adopted. This process supports substrate PNP transistors, which could be used in the temperature compensation scheme of the reference. The scaled frequency reference implemented in this process [5], utilizes an ETF, which is about $5\times$ smaller than the one in previous work [6]. This is in line with the expected $4.5\times$ improvement of lithographic inaccuracy in comparison with the 0.7 μm process. After this introductory section, the chapter progresses with a description of the scaling strategy. Furthermore, the effect of scaling on the circuit and system design of the electrothermal frequency reference will be discussed. Experimental results on the test chip will be provided and the chapter ends with conclusions.

6.2 Scaling Strategy

As described in Chaps. 4 and 5, an electrothermal frequency reference can be realized by embedding an ETF in a digitally-assisted frequency-locked loop (DAFLL). In such a loop the output frequency of a DCO, f_{DCO}, is locked to the ETF phase shift, ϕ_{ETF} (see Fig. 6.1), by means of feedback. This ensures that the DCO operates at a frequency, where $\phi_{ETF} = \phi_{ref}$, where ϕ_{ref} is a temperature dependent phase reference. This reference is provided by an on-chip, band-gap temperature sensor, which ensures that $\phi_{ref} \propto T^{0.9}$, the same temperature dependence as ϕ_{ETF} (see Chap. 5). For the simple point-heater point-temperature sensor model of ETF shown in Fig. 6.1, the ETF phase ϕ_{ETF} is determined by the heater to temperature sensor distance, denoted by s (see Chap. 3).

The chosen scaling strategy aims to maintain the inaccuracy of the electrothermal frequency reference at the $\pm 0.1\%$ achieved in [6], while improving other performance metrics. From (3.9), we can conclude that:

$$\phi_{ETF} \propto s\sqrt{f_{DCO}/D}. \tag{6.1}$$

where D is the thermal diffusivity of silicon. From (6.1) it can be seen that scaling an ETF, i.e. decreasing s, implies that for a given ϕ_{ETF} the ETF can be operated at a

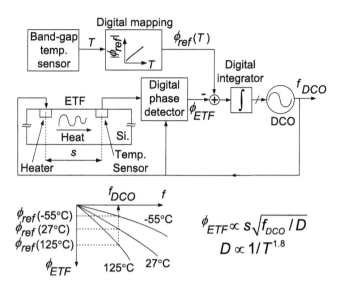

Fig. 6.1 Simplified block-diagram of an electrothermal frequency reference (*top*), and the phase frequency characteristic of an ETF at various temperatures (*bottom*)

higher frequency. However, decreasing s also results in greater sensitivity to lithographic errors. By differentiating (6.1) we obtain:

$$\frac{\Delta \phi_{ETF}}{\phi_{ETF}} = \frac{\Delta s}{s}. \tag{6.2}$$

which implies that for a given process, and thus a given lithographic error Δs, the relative accuracy of ϕ_{ETF} decreases as s decreases [7]. However, increasing s also reduces the thermopile's output amplitude, which for $s = 24 \ \mu m$ and 2.5 mW of heater power dissipation is only a few hundred micro-volts [6]. Since this is quite a small signal, the thermopile's thermal noise will be a major contributor to the DAFLL's jitter. This can only be mitigated by increasing the ETF's heater power dissipation. The choice of s thus involves a fundamental trade-off between accuracy, output frequency, jitter, and power dissipation (see Chap. 3).

To break this trade-off, a more advanced process can be used, thus reducing Δs and allowing the ETF to be scaled without losing accuracy. In the scaled electrothermal frequency reference presented in this chapter, a 0.16 μm CMOS process has been adopted. To exploit its improved lithographic accuracy with respect to the 0.7 μm process used in [6], s has been scaled by about 5×, i.e. reduced from 24 to 4.7 μm.

After fixing s, the heater and the thermopile of the scaled ETF need to be designed. A minimum size U-shaped heater based on a p^+ diffusion resistor has been adopted (see Fig. 6.2 for the heater dimensions). Based on this geometry, the heater has a resistance of $R_{heater} \sim 800\Omega$, which allows for a maximum power dissipation of 2 mW, considering the 1.8 V supply. The combination of the design rules and the desired s meant that a maximum of 16 thermocouple arms could be

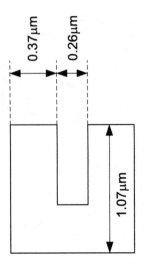

Fig. 6.2 The p⁺ heater dimensions of the scaled ETF

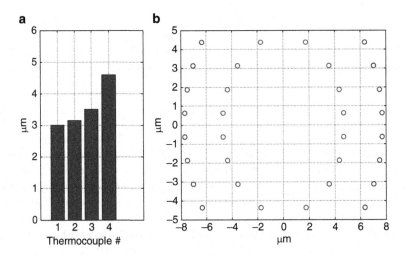

Fig. 6.3 (**a**) The thermocouple arms per thermopile quadrant for the scaled ETF, and (**b**) the layout of the ETF illustrating the position of the hot and cold thermopile junctions

fitted around this heater. These are placed in 4 quadrants, each made of 4 arms. This allows for a differential structure made of two half-thermopiles, laid out in crossed couples.

To maximize the ETF's efficiency, the thermopile's *hot* junctions are situated around the heater on a constant phase-shift contour (see Chap. 3). The location of its *cold* junctions was then optimized for maximum output and minimum resistance (Chap. 3, Sect. 3.4.2). This resulted in four different thermocouple arm lengths for each of the thermopile quadrants, which is shown in Fig. 6.3a. The locations of the

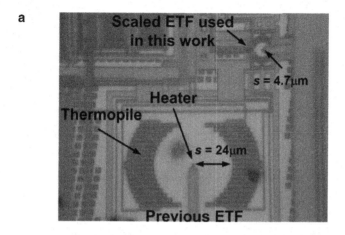

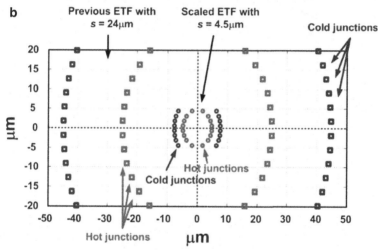

Fig. 6.4 (a) Photomicrograph of the previous ETF and the scaled ETF; (b) Locations of the hot and cold junction of the corresponding thermopiles

hot and *cold* thermopile junctions of the scaled ETF are shown in Fig. 6.3b. A photomicrograph of the scaled ETF adjacent to the previous ETF is shown in Fig. 6.4a. The locations of their thermopile junctions are compared in Fig. 6.4b.

Apart from optimizing the location of the thermopile's junctions, the signal-to-noise ratio SNR_{ETF} at the output of an ETF can also be improved by optimizing the number of thermocouples n_{tp} in the thermopile. The ETF's output signal level will be determined by n_{tp}, the temperature gradient across the thermopile and the Seebeck coefficient of each thermocouple, while its thermal noise will be determined by its total resistance R_{tp}. SNR_{ETF} may thus be expressed as [7]:

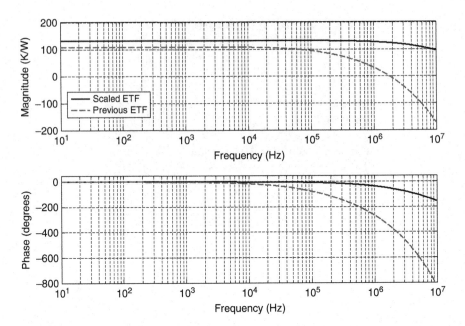

Fig. 6.5 Thermal impedance simulations at room temperature, comparing amplitude and phase response of scaled ETF with previous ETF

$$SNR_{ETF} \propto \frac{n_{tp}}{s\sqrt{R_{tp}}}. \tag{6.3}$$

Since the choice of the *hot* and *cold* junctions fixes the area of the thermopile, $R_{tp} \propto n_{tp}$ and so SNR_{ETF} will be fixed [8]. However, the output level can be maximized by maximizing n_{tp}, subject to the design rules of the chosen process. The resulting scaled ETF has fewer arms than the previous design (16 vs. 24), lower resistance (8kΩ vs. 20kΩ) and a 5.5× greater SNR:

$$\frac{SNR_{ETF,new}}{SNR_{ETF,old}} = \frac{\frac{n_{tp,new}}{s_{new} \cdot \sqrt{R_{tp,new}}}}{\frac{n_{tp,old}}{s_{old} \cdot \sqrt{R_{tp,old}}}} = \frac{\frac{16}{4.7\mu \cdot \sqrt{8k}}}{\frac{24}{24\mu \cdot \sqrt{20k}}} \approx 5.5. \tag{6.4}$$

From (6.1), it can be seen that for a given ϕ_{ETF}, scaling s by a factor k will scale its driving frequency f_{DCO} by a factor k^2. Compared to the previous ETF, which was operated at 100 kHz, a 5× scaled ETF should thus be operated at about 3 MHz for the same ϕ_{ETF}. This was confirmed by numerical simulations of the thermal impedance (the relationship between heater power and sensor temperature) of both ETFs at room temperature (Fig. 6.5). However, since ϕ_{ETF} cannot be distinguished from the process-dependent phase-shift of its readout circuit, increasing the drive frequency would require a proportional increase in the bandwidth of the

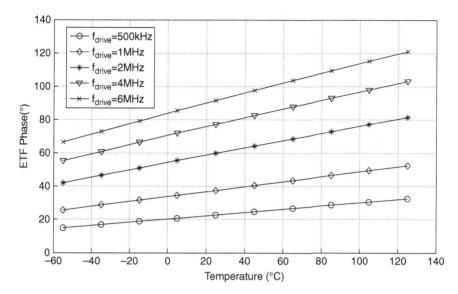

Fig. 6.6 The expected phase response of the scaled ETF as a function of temperature and the driving frequency

readout circuit. As a compromise, it was decided to drive the scaled ETF at 1 MHz. At this frequency, ϕ_{ETF} will vary from 40° to 70° over the military temperature range (Fig. 6.6).

The characteristics of the scaled ETF affect the design of the rest of the electrothermal frequency reference. Compared to [6], the bandwidth of the ETF's readout circuitry must be increased by 10× in order to maintain the same level of phase accuracy. Also, to benefit from the 5.5× improvement in SNR, the resolution of both the DCO and the band-gap temperature sensor must be improved. The design of these blocks will be the topic of the next sections.

6.3   System-Level Design

A simplified block-diagram of the scaled electrothermal frequency reference is shown in Fig. 6.7. It uses the same architecture as the one described in the previous chapter. Compared to previous work, the resolution of D$\Delta\Sigma$M has been increased from 12 to 15 bits. Furthermore, for the 30° expected phase variation of the scaled ETF (see Fig. 6.6) over the military temperature range, the modulator's resolution will be about 1 m°, and so should not limit the trimming resolution of the reference.

The ETF's phase shift, ϕ_{ETF}, is digitized by the PD$\Delta\Sigma$M with reference to its two phase references, $f_{\text{drive}}(\phi_0)$ and $f_{\text{drive}}(\phi_1)$, which are digitally phase-shifted versions of f_{drive} (see Chaps. 4 and 5). In this design, $f_{\text{DCO}} = 16 \cdot f_{\text{drive}} = 16$ MHz, which means that $f_{\text{drive}}(\phi_0)$ and f_{drive} (ϕ_1) can be chosen in steps of 22.5°.

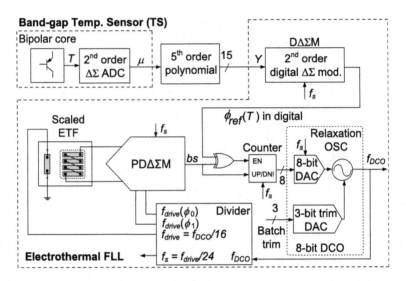

Fig. 6.7 Simplified block-diagram of the scaled electrothermal frequency reference

The modulator's bitstream output, bs, will then be a digital representation of ϕ_{ETF}. The bitstream difference between ϕ_{ETF} and ϕ_{ref} is integrated by a 12-bit up/down counter, which suppresses the quantization noise of both $\Delta\Sigma$ modulators. The counter is incremented or decremented according to the polarity of the bitstream difference and is disabled, by an XOR gate, when the bitstreams are equal. The counter's eight MSB's then update the DCO at a sampling rate of $f_s = f_{drive}/24$ = 41.66 kHz.

The DCO consists of an RC oscillator that is driven by an 8-bit fine DAC. This, in turn, produces a tuning current that results in a 0.02% LSB when applied to the oscillator. This is in line with the expected accuracy and is more than $4\times$ lower than the expected jitter. A 3-bit coarse DAC is used to compensate for the effect of process spread on the DCO's center frequency. Compared to [6], which used a 12-bit fine DAC, the use of a temperature-compensated RC oscillator in this design means that only an 8-bit fine DAC is required. This could then be implemented with a straightforward 256-element resistor ladder, while also guaranteeing the monotonicity required for loop stability.

6.4 Error Sources

The major sources of error in the scaled electrothermal frequency reference are the electrical phase shift associated with the finite bandwidth of the PD$\Delta\Sigma$M, the input-referred offset of the PD$\Delta\Sigma$M, and the temperature sensing inaccuracy of TS. With the help of the system model shown in Fig. 6.7, the effects of these errors have been simulated at three temperature points: $-55°C$, $27°C$, and $125°C$. The finite bandwidth of the PDΔΣM will give rise to process-dependent electrical phase-shift which will be indistinguishable from the ETF's own thermal phase-shift. As shown in Fig. 6.8a, the worst-case error occurs at −55°C, when a phase error of only 20 milli-degrees causes a frequency error of 0.1%. To ensure that its error contribution is much less than 0.1%, the bandwidth of the PDΔΣM's front-end was designed to be in excess of 500 MHz.

The residual offset of the PDΔΣM is also an important source of error (Chap. 4). The PDΔΣM employs a chopper demodulator as a synchronous phase detector. The chopper's switching action and charge injection mismatch is a source of residual offset [9], which adds error to the demodulated phase and thus leads to errors in the DAFLL's output frequency (see Chap. 4). As shown in Fig. 6.8b, an input-referred residual offset of 2 μV will cause a frequency error of about 0.05% at 125°C. It should be noted that the increased frequency of operation from 100 kHz to 1 MHz will increase the residual offset proportionally [9, 10]; however, the technology scaling, and thus the use of smaller size switches, should reduce the mismatched charge injection (see Chap. 4). Furthermore, the required offset performance can be easily achieved by applying low-frequency chopping, i.e. by chopping the drive signals applied to the ETF as well as the digital output of the PDΔΣM.

The effect of TS inaccuracy on the output frequency is shown in Fig. 6.8c. It can be seen that a temperature sensing error of only 0.1°C results in a frequency error of about 0.08%. Since an inaccuracy of 0.1°C reflects the state-of-the-art [11, 12] this represents a fundamental limit on the accuracy of electrothermal (thermal-diffusivity-based) frequency references. Further improvements will require more accurate temperature sensors or the use of multi-point (e.g. two temperature points) trimming.

6.5 Circuit Realizations

The following sub-sections describe the design of the major analog building blocks of the scaled electrothermal frequency reference: the ETF and its drive circuitry, the phase-domain ΔΣ modulator (PDΔΣM), the DCO, and the band-gap temperature sensor (TS).

6.5.1 ETF and PDΔΣM

6.5.1.1 Detailed System Overview

The heater of the scaled ETF is driven at a frequency $f_{drive} = f_{DCO}/16 = 1$ MHz, by a heater drive circuit (HD) (see Fig. 6.9a). The ETF's phase shift ϕ_{ETF} is then digitized by the PDΔΣM, whose sampling clock and phase reference signals are also derived from f_{DCO}. At a sampling rate of $f_s = 41.66$ kHz, the DAFLL's noise-bandwidth is only about 10 Hz. In this bandwidth, the modulator's resolution

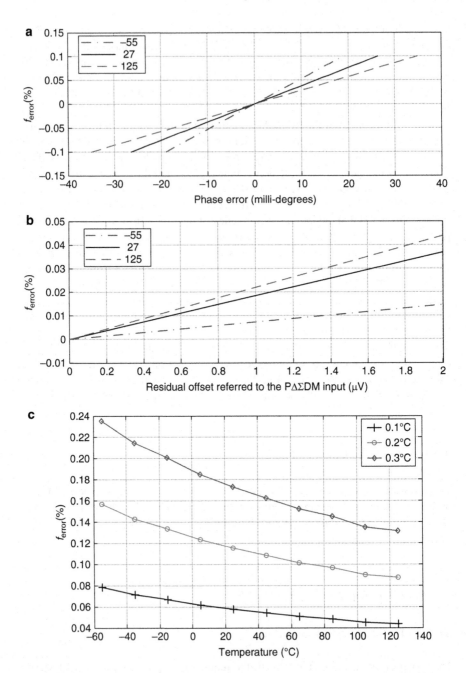

Fig. 6.8 System-level simulations of the scaled electrothermal frequency reference over temperature showing frequency error as a function of: (**a**) excessive phase error due to the limited bandwidth of PDΔΣM front-end; (**b**) residual offset error referred to the PDΔΣM input; (**c**) a fixed absolute temperature measurement error made by TS

Fig. 6.9 PDΔΣM: (a) simplified circuit diagram including ETF and its heater driving circuitry; (b) timing diagram of phase references and ETF drive

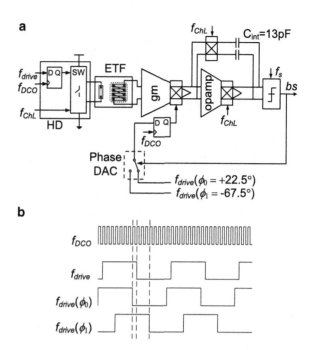

should be limited by the ETF's thermal noise (12 nV/√Hz). For the expected 30° variation in ϕ_{ETF} over the military temperature range, this noise level translates into a resolution of about 5 milli-degrees. Since the ΔΣ modulator's over-sampling ratio is greater than 4,096, a first-order modulator will be sufficient to ensure that its quantization noise is below the ETF's thermal noise [13].

The PDΔΣM (Fig. 6.9a) consists of a gain-boosted folded-cascode transconductor (gm) that converts the ETF's output into a current. This is then phase detected by a chopper demodulator, which multiplies it by f_{drive} ($\phi_0 = +22.5°$) or f_{drive} ($\phi_1 = -67.5°$), provided by a phase DAC and depending on the value of the bitstream, bs (Fig. 6.9b). The residual offset of the chopper demodulator is suppressed by locating it within the gain-booster loop of the transconductor's cascode transistors (Chap. 4). The chopped output current, whose DC value is proportional to the difference between ϕ_{ETF} and the phase DAC's phase output, is then applied to an active integrator built around a two-stage opamp ($C_{int} = 13$ pF). This suits the 1.8 V supply much better than the G_m-C integrator used in previous work [6]. The opamp's DC gain is > 100 dB, which ensures that the quantization noise floor established by integrator leakage is still well below the thermal noise level [13]. To further reduce the demodulator's residual offset, the entire front end (including the ETF) is low-frequency chopped at $f_{ChL} = f_{drive}/16,376 = 61.06$ Hz.

At 1 MHz, the PDΔΣM's simulated phase error (including the influence of the thermopile's parasitic capacitance) is less than 7 milli-degrees over process and temperature, which leads to a frequency error of less than 0.025%. The modulator consumes 290 μA, which is about $2\times$ less than [6] and has an input referred noise of 10 nV/√Hz. About 70% of its total power is consumed by the OTA while the rest is consumed by the opamp and the comparator.

Additional phase delay between the ETF's heater drive and the demodulator's phase references will increase the error in ϕ_{ETF}. For the scaled ETF, a 200 ps delay at $f_{drive} = 1$ MHz leads to an output frequency error of 0.25%. To suppress this effect, the clock-synchronized heater driving block (HD) has been added. As shown in Fig. 6.9a, the use of D flip-flops clocked at f_{DCO} ensures that the additional phase delay remains well below 15 ps over the process and temperature variations.

The detailed transistor-level design of the PDΔΣM's various building blocks, i.e. its transconductor, active integrator, and quantizer as well as a detailed design of the ETF heater drive circuitry (HD) will be described next.

6.5.1.2 Transconductor

The 1 MHz drive frequency of the scaled ETF is $10\times$ larger than the previous work, which requires special attention to the design of the transconductor. The main concern is the excessive electrical phase spread at the excitation frequency of 1 MHz resulting in a degradation of the reference's accuracy. To increase safety margins, it was decided to keep the option for having an equivalent of the previous ETF ($s = 24$ μm) implemented into the design.

The previous ETF's 100 kHz drive frequency is much less sensitive to excess phase errors; however, it imposes other challenges due to the larger chopper ripple amplitudes. Such ripple is due to the offset of the transconductor, which is modulated to f_{drive} by the synchronous demodulator. In the previous design, the 100 kHz ripple over the PDΔΣM's integrator had an amplitude of about 500 mV. This is not a very important issue for a 5 V process (0.7 μm), however for a 1.8 V process (0.16 μm), such large ripple could waste a lot of signal headroom, and eventually result in clipping of the integrator. Therefore, the possibility of auto-zeroing the transconductor in order to reduce its initial offset has been considered. It should be noted that the options mentioned above are to add margin and thus should not be necessary in case the scaled ETF performs as expected at 1 MHz.

To develop insight into the expected internal signal of the PDΔΣM, i.e. the chopper ripple and the maximum expected signal swing on its integrator, a series of simulations were performed on the circuit of Fig. 6.9a. The summary of simulation results are provided in Table 6.1. This table reports the largest swing and ripple over the integrator for different sampling frequencies. These simulations consider a 1 mV ETF output signal at both 1 MHz and 100 kHz, and assume the largest phase input for the modulator (the full-scale phase signal). A 5 mV offset has been considered for both the transconductor (gm) and the integrator opamp. A comparison of these results for the auto-zeroed (denoted by 'W AZ') and non auto-zeroed (denoted by 'no AZ') cases shows that the operation of the system with the scaled ETF and an $f_{drive} = 1$ MHz should not require auto-zeroing.

Table 6.1 The simulated integrator swing and the ripple due to the offset chopped by the demodulator at different f_{drive} and f_s values

ETF f_{drive}	f_s	Largest integrator swing (No offset)	Largest integrator swing (gm offset = 5 mV and opamp offset = 5 mV)	High frequency ripple caused by the offset chopped by the synch. demodulator
1 MHz	500 kHz	40 mV	150 mV	100 mV
1 MHz	100 kHz	200 mV	300 mV	100 mV
1 MHz	20 kHz	1 V	1.1 V	100 mV
100 kHz	50 kHz	400 mV	No AZ: 1.1 V – W AZ: 450 mV	No AZ: 800 mV
100 kHz	25 kHz	800 mV	No AZ: clips – W AZ: 800 mV	No AZ: 800 mV
100 kHz	12.5 kHz	1.6 V	No AZ: clips – W AZ: 1.6 V	No AZ: 800 mV

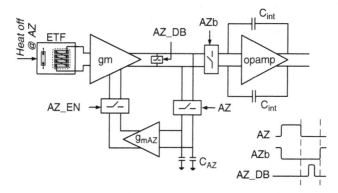

Fig. 6.10   The auto-zeroing circuitry associated with the transconductor and its timing

Figure 6.10 shows the simplified auto-zeroing circuitry of the transconductor (gm) with its switches and associated timing diagram [10]. Since for the expected operating conditions with the scaled ETF involved, no auto zeroing is expected to be used, an enabling signal *AZ_EN* can be used to disable the auto zeroing loop completely. If the transconductor needs to be auto-zeroed, *AZ_EN* will be asserted, and the ETF heater drive will be disabled to remove the signal at the input of gm, which effectively short circuits its input.

To perform auto-zeroing, the gm's output will be connected to the input capacitors of an auxiliary transconductance g_mAZ via the switches driven by *AZ*. Meanwhile, *AZb* switches disconnect the integrator from gm. The feedback loop formed by gm and g_mAZ (assuming zero input for gm) causes the offset of gm to generate a current that will be integrated by C_{AZ}. The auxiliary g_mAZ will then drive a correction current into a node inside gm, which cancels the initial offset (this will be shown later). The loop gain determines the final level of offset reduction [9]. Once auto zeroing is completed, the parasitic capacitances at the output of gm are discharged through dead-banding switches *AZ_DB*. This is done before the output of gm is re-connected back to the integrator input by *AZb*. This minimizes the voltage spikes that occur when the integrator is re-connected to gm.

A complete circuit diagram of gm is shown in Fig. 6.11. It is basically a folded-cascode OTA with PMOS inputs. A current multiplexer, MUX, steers a tail current to any of the desired four input pairs. Each input pair can then be connected to a different ETF test structure. The 60 μA tail current results in an input transconductance of 570μS within the input devices with $W/L = 60$ μm/0.36 μm. This relatively small ratio was chosen to reduce their gate capacitance, which contributes to the electrical filtering of the ETF signal. This size of input pair in combination with the magnitude of the tail current means that the devices operate in strong inversion. The input common mode, i.e. the bias voltage of the ETF thermopile, was set to 400 mV.

The transconductor's output common mode of 900 mV is regulated by CMFB circuitry, as in the previous design (Sect. 4.3.2.1). The CMFB loop is frequency compensated by 1.5 pF capacitors and has a phase margin of 67°. To minimize the residual offset due to the switching action of the chopper demodulators, they are located at the virtual grounds provided by the booster amplifiers. The PMOS and NMOS gain-boosters (Fig. 6.12) consume 10 μA and 16 μA respectively and provide DC gains in the order of 60 dB. The gain-boosting loops at the PMOS and NMOS sides are frequency compensated by 800fF capacitors (Fig. 6.12) for 60° loop phase margins.

The transconductor of Fig. 6.11 has a simulated nominal DC gain of 140 dB, which can vary from 110 to 145 dB over the process, temperature and mismatch. Its estimated input-referred offset is about 2.5 mV, and its maximum residual offset current when the chopper demodulators are operated at 1 MHz, is about 3 nA (about 5 μV input-referred). The simulated maximum excess phase shift added to the ETF signal before phase detection amounts to 17 milli-degrees over temperature, process, and mismatch (at 1 MHz). Within one batch, this can be as low as 3 milli-degrees. Considering the parasitic capacitance associated with the scaled ETF's thermopile, the excess phase shift can amount to about 7 milli-degrees. The transconductor consumes a total supply current of 210 μA. The total modulator input-referred noise (including the opamp noise) is about 10 nV/√Hz.

6.5.1.3 Integrator Opamp

A detailed schematic of the integrator opamp is shown in Fig. 6.13. It is a two-stage amplifier with a telescopic first stage, PMOS input pair, and NMOS common-source output stages. At a total supply current of 42 μA, it has a nominal simulated DC gain of 110 dB (97–119 dB over the process and temperature). With a Miller capacitor of $C_m = 4$ pF and for the total load capacitance of the integrator, the amplifier has a unity gain bandwidth of about 1 MHz with a phase margin of 62°.

The opamp's input chopper is merged with the feedback path (see Fig. 6.9 and [14]), and the output chopper is located before the output stage. This will not

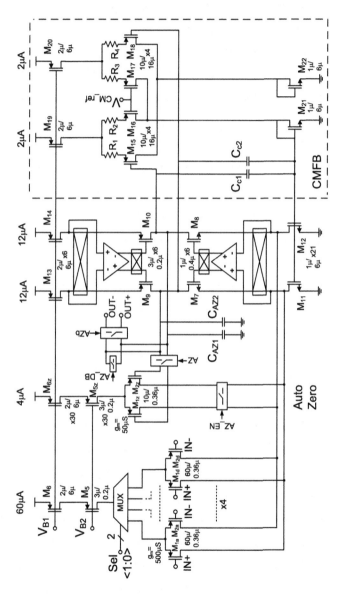

Fig. 6.11 The detailed circuit diagram of the PDAΣM's front-end transconductor (gm)

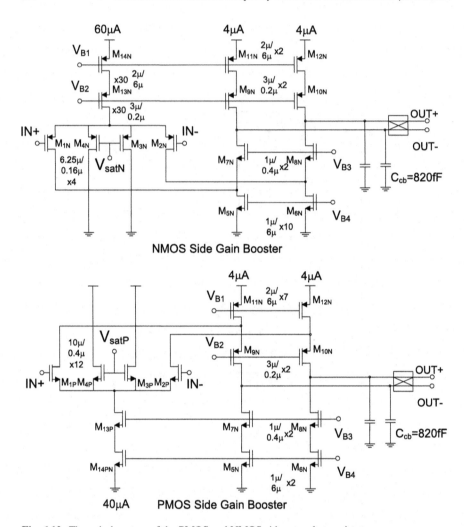

Fig. 6.12 The gain-boosters of the PMOS and NMOS side cascode transistors

increase the residual offset because the total input-referred offset of the output stage is suppressed by the input-stage gain. The total estimated opamp input-referred offset is about 3.5 mV (non-chopped). The amplifier's fully differential output swing is ±3 V over process and temperature. Resistors $R_{CMFB} = 300k\Omega$ provide an estimate of the output common mode level, which then the CMFB loop including M_{16-18} regulates to the common mode reference, V_{cm_ref} [9].

6.5.1.4   Comparator

The PDΔΣM's comparator (Fig. 6.14) was designed to be operational for sampling frequencies as high as $f_s = f_{drive}/2 = 500$ kHz. This includes a pre-amplifier with a

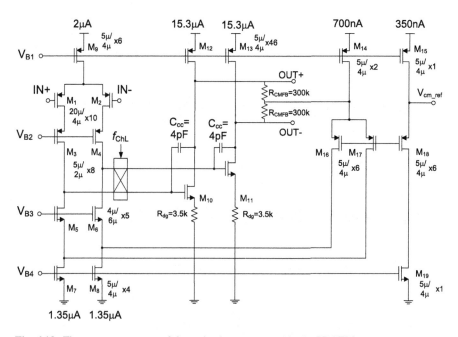

Fig. 6.13 The two-stage opamp of the active integrator used in the PDΔΣM

gain of 8 to reduce the *kick-back* effect of the following positive feedback latch. A minimum-size reset switch M_{rst} driven by the comparator clock signal *PhEval*, releases the latch from reset state at the sampling moment of the modulator. When the switch opens, the latch regenerates and its output drives a set-reset flip-flop via output stage made by transistors M_{11-14}. The comparator's supply current is 16 μA, and has a maximum input-referred offset of 1 mV over process and temperature. At sampling rate of 500 kHz and with an input common mode of 900 mV its minimum input signal is $500\mu V_{pk-pk}$.

6.5.2   DCO

The DCO is an RC relaxation oscillator, whose output frequency is tuned by an 8-bit fine DAC (see Fig. 6.15). Compared to the previous implementation, the DAC employs a simpler and smaller unary architecture in order to guarantee monotonicity. The effect of process spread is then corrected by means of a 3-bit batch trim applied by means of a coarse DAC. The oscillator includes comparators $U_{1,2}$ which compare $V_{1,2}$ to V_{ref}. While one capacitor is being charged through R_{ch} to V_{DD}, the other one is being discharged by I_{ref} [15]. The speed at which the capacitor is charged has been reduced via resistor R_{ch}. This is to minimize the current spikes drawn from the analog supply and eventually injected into the analog ground when

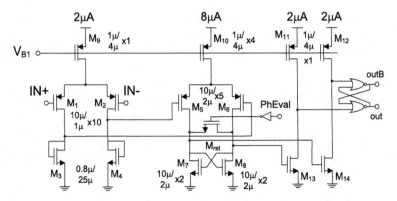

Fig. 6.14 Schematic of the comparator used in the PDΔΣM

the capacitor is connected to V_{DD}. The states of the SR-latch and the chopper CH1 change when the voltage across the capacitor reaches V_{ref}.

The nominal f_{DCO} is 16 MHz, which can be tuned to a range of ± 500 kHz through the fine 8-bit DAC. This is enough to correct for the mismatch and temperature variation effects. The batch trim, which is applied by the coarse DAC, has a range of ± 3 MHz and can be adjusted with steps of 750 kHz. For this reason, V_{ref} is varied using a 3-bit DAC made of the divider $R_{t0..8}$, switches $S_{t0..7}$, and the buffer U_3. The batch trim is a process corner trim and thus will be fixed for the entire batch. During the normal operation of the DFLL, the loop constantly tunes the DCO by tuning I_{ref} through the fine DAC.

The 8-bit input code to the DCO selects only the nth switch from the switches $S_{0..255}$ copying V_A to the corresponding point on the resistive ladder made of unit resistors $R_{0..255} = R = 125\Omega$. This produces a current I_{ref}:

$$I_{ref} \propto \frac{V_{DD} - V_A}{R_A + (n \times R)}. \tag{6.5}$$

Since there is no current through the DAC switches, their on-resistance will be of no importance allowing for the use of minimum-size devices. The 256 elements of the resistive ladder are arranged in a 16×16 matrix with row and column decoders made of two binary-to-thermometer decoders. The inputs to the decoders are the 4 MSB and LSB bits of the DAC 8-bit binary input. This also allows for a matrix layout of the DAC elements, each including a resistor R and switch S_x.

The DCO is further temperature-compensated by adding an N-well resistor R_N with positive TC to the divider, generating V_A. The temperature compensation and batch trim ensure that the range of the 8-bit fine DAC is enough to compensate for the VCO's residual spread with a step size of 200 ppm. Since both V_A and V_{ref} are supply-referenced, the DCO output frequency is insensitive to supply-voltage variations.

The circuitry of the oscillator's comparators resembles those used in the previous design. The comparators' -3 dB bandwidth is 100 MHz and their supply current is 50 μA. A 50 mV hysteresis is built into the comparators and their input

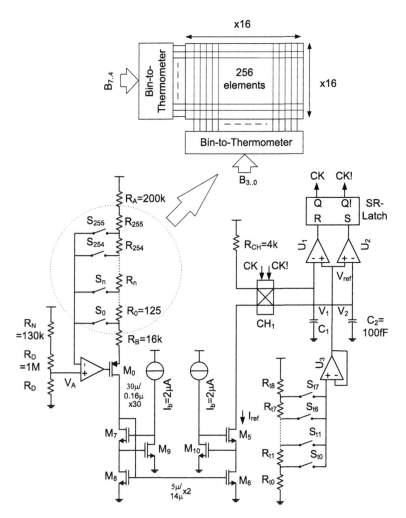

Fig. 6.15 Detailed schematic of the DCO including the 8-bit fine DAC (showing the arrangement of the 256 DAC elements), the oscillator, and the 3-bit coarse DAC

referred noise is 150 μV (rms). This noise and the $1/f$ noise due to the current I_{ref} result in a simulated cycle-to-cycle jitter of 20 ps (rms) at a nominal output frequency of 16 MHz. The DCO's total supply current is 117 μA, of which 100 μA is consumed by the oscillator and the rest by the two DACs.

6.5.3   Temperature Sensor

A band-gap temperature sensor (TS), is used to temperature compensate the DAFLL. This includes a bipolar core and a ΔΣ ADC [11, 12]. The bipolar core

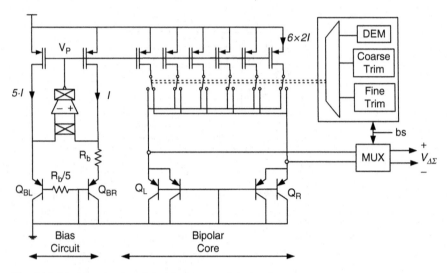

Fig. 6.16 The bias circuit and the bipolar core of the band-gap temperature sensor

produces a PTAT voltage ΔV_{BE}, which is the difference between the base–emitter voltages of two substrate PNP transistors. These are biased at a different current ratio. This is then combined with a trimmable CTAT voltage V_{BE}, which is the base–emitter voltage of only one of those PNP devices. The PTAT and CTAT values are fed to a charge balancing $\Delta\Sigma$ modulator that produces a ratiometric value μ defined by (5.9). This digital number is a representative of the die temperature (see Fig. 5.8).

The bipolar core of TS has been adapted from a micro-power temperature sensor design implemented in the same 0.16 μm process [16, 17]. This design uses a fast two-step or a zoom ADC (the combination of a SAR algorithm and a $\Delta\Sigma$ data converter) [16, 17]. Considering the required resolution of less than 10 mK, a second order charge balancing $\Delta\Sigma$ modulator was designed to be interfaced with the abovementioned bipolar core. The ADC is based on the single-capacitor architecture introduced in Sect. 5.3.2.2. Since the system-level design and accuracy requirements of TS resemble the previous one (Chap. 5), only the circuit realizations will be briefly discussed here.

Figure 6.16 illustrates a schematic diagram of the bipolar core [17]. A PTAT bias generator on the left side produces a PTAT bias current I via two PNP transistors (5×5 μm) that are biased at a 5:1 current ratio. With the variations in β_F of the bipolar devices (over process and temperature), a β_F compensation circuitry, including a resistor $R_b/5$ in series with the base of Q_{BL}, is incorporated [11]. Due to feedback, the collector current of the bipolar transistors in the bias circuit will therefore be independent from β_F of the PNPs, and thus the resulting V_{BE} produced in the bipolar core will not be altered by this non-ideality (see Sect. 5.3.2.2). The value of I is chosen to be 90 nA at room temperature. This ensures that both I and $5 \cdot I$ are in a relatively flat region of the PNP's β_F versus bias current characteristic.

The opamp has been chopped to reduce the temperature sensing inaccuracy induced by its offset [11, 17].

The current I produced by the bias circuit is mirrored to six current sources with a gain of 2. This means that each current source carries a PTAT current $2 \times I$ (180 nA) at room temperature. Depending on the bitstream polarity of the charge balancing $\Delta\Sigma$ modulator, the bipolar core produces V_{BE} or ΔV_{BE} as input to the converter (Sect. 5.3.2.2).

If ΔV_{BE} needs to be applied to the modulator, a 5:1 current ratio is applied to both Q_L and Q_R, respectively. The accuracy of the current sources determines the accuracy of the PTAT voltage ΔV_{BE}. As in the previous design a dynamic element matching cycle rotates the unit current source that determines the 5:1 ratio at the ΔV_{BE} phase.

If V_{BE} has to be the input to the modulator, a coarse trim can be applied to the bias current of the transistor Q_L or Q_R that produces V_{BE}. This can be done by selecting a range of one to five of the current sources. In this phase, the sixth current source is modulated by a digital $\Delta\Sigma$ modulator in order to apply a fine trim to the V_{BE} (see Sect. 5.3.2.2). A multiplexer MUX at the output of the bipolar core determines whether the ΔV_{BE} or V_{BE} will be input to the $\Delta\Sigma$ modulator (controlled by the modulator's bitstream bs).

The operating principle of the second-order charge balancing $\Delta\Sigma$ modulator of TS exactly resembles that of the previous design. In the switched-capacitor modulator (Fig. 6.17), the PTAT gain $\alpha = 16$ is generated by sampling ΔV_{BE} 16 times on a unit sampling capacitor. The modulator utilizes the same feed-forward topology as the previous one and its coefficients remain the same; however, the capacitor values are scaled. The scaling of the capacitors is referred to the scaling of the sampling capacitor, which determines the modulator's input-referred thermal noise. A thermal noise limited resolution better than 10mK can be achieved by a 500 fF input capacitor at $\alpha = 16$ and for 1,048 samples within one incremental conversion (corresponding to a conversion rate of 2 Hz) [11]. The modulator coefficients as well as the other capacitor values are:

$$a_1 = \frac{C_S}{C_{int1}} = \frac{500fF}{2pF} = \frac{1}{4}, a_2 = \frac{C_F}{C_{int2}} = \frac{200fF}{400fF} = \frac{1}{2}, b = \frac{C_{FF}}{C_F} = \frac{100fF}{200fF} = \frac{1}{2}. \quad (6.6)$$

Apart from a few changes in the modulator timing (shown in Fig. 6.18), the only major change (compared to the previous design) is the topology of the second integrator. This was modified from the previous non-auto zero structure to an auto-zeroed version. This is just for compatibility with the different timing used in the switched-capacitor common-mode feedback of the integrator opamps. This helps reduce the offset and $1/f$ noise of the second integrator. The opamps used in the first and second integrator are the same. They utilize a conventional folded-cascode topology with a quiescent current of 1.66 μA and provide a DC gain of 90 dB with a phase margin of 62° at a unity gain bandwidth of 1.8 MHz.

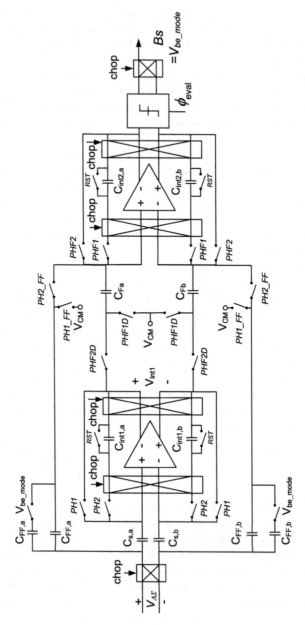

Fig. 6.17 The second-order charge- balancing switched-capacitor ΔΣ modulator of the temperature sensor with both auto-zeroing integrators

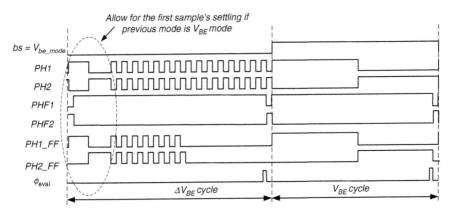

Fig. 6.18 Timing diagram of the delta sigma modulator for the ΔV_{BE} and V_{BE} cycles

The modulator's sampling frequency is 2 kHz. As the timing diagram in Fig. 6.18 shows, the first ΔV_{BE} sampling phase, which directly follows a V_{BE} phase, has been given extra time to allow for better settling of the sampling capacitor. This is mainly because after a V_{BE} phase, one of the capacitors should always be charged from ground, which requires more settling time (described in Chap. 5). With 500fF sampling capacitor, and at the aforementioned operating frequency, the settling of the bipolar core when sampling ΔV_{BE} is better than 10mK ($16\times$ higher sampling rate in this phase compared to the modulator's sampling frequency).

System level chopping has been applied to the modulator (see Fig. 6.17) to remove the residual offset associated with the mismatched charge injection of the switches. The system level chopping is performed in a bitstream-controlled manner [12] in order to break the correlation between the tones associated with the chopping ripple and the modulator's quantization noise. This avoids folding of the quantization noise to the band, which can cause an increase in the modulator's noise floor. The bitstream controlled chopping mechanism changes the chopper state only when the bitstream polarity is high. A total of two system level chopping periods are accommodated within every conversion of the TS.

Compared to the previous implementation (see Fig. 5.13) the timing associated with the modulator's feed-forward capacitor C_{FF} is also modified. The capacitor is not discharged to the second integrator's integrating capacitor during the auto-zeroing phase. This was not a problem in the previous implementation when a non auto-zero second integrator was used. The new timing of the TS ensures that the complete charge of C_{FF} is always transferred during the first half of the modulator cycle (see Fig. 6.18). This means that the effective amount of charge transferred during the ΔV_{BE} cycle is halved. To compensate for this, a second C_{FF} capacitor will be connected in parallel with the main C_{FF} capacitor during this phase (see Fig. 6.18).

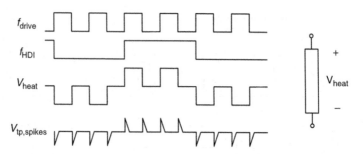

Fig. 6.19 The heater drive inversion (HDI) applied to the heater drive signal to reverse the polarity of the capacitive-coupled spikes

6.5.4   Heater Drive Circuitry

On-chip heater drive circuitry has been implemented in the scaled electrothermal frequency reference. In the previous 0.7 μm design, the heater driver switches were implemented by means of 5Ω analog off-chip switches realized by ADG719 devices [18]. The main task of the heater drive circuitry is to provide the required current to the ETF heater with a minimum ohmic dissipation and without the addition of excess phase delay. Implementing an on-chip heater drive allows for synchronization of the heater drive signal with the phase references of the PDΔΣM.

With the 10× increase of the drive frequency (1 MHz compared to the previous 100 kHz), synchronization is rather crucial. This is because any additional phase delay between the ETF's heater drive and the demodulator's phase references will increase the error in ϕ_{ETF}. For the scaled ETF, a 200 ps delay leads to an output frequency error of 0.25%.

Furthermore, the heater drive circuitry needs to be able to drive the heater with a bi-directional current. This is to implement a heater drive inversion scheme (HDI) [7] in order to cancel the capacitively coupled spikes appearing at the ETF's thermopile. These spikes, (Fig. 6.19) are in phase with the heater drive and thus cause a DC error at the output of the PDΔΣM's synchronous demodulator. The error induced by them is more for the scaled ETF, where the heater thermopile distance is reduced to 4.7 μm. As described in [7], this problem can be solved by periodically inverting the polarity of the heater drive at a rate of $f_{drive}/2$ (see Fig. 6.19). This emulates an effective chopping applied to the resulting spikes' polarity. Furthermore, the low-frequency chopping of the PDΔΣM requires the heat signal in the ETF to be chopped as well. This requires that the heater drive circuitry changes the heat generation phase by 180°, synchronous to the low frequency chopping signal f_{ChL}.

The heater drive circuit includes a bridge circuit (see Fig. 6.20). This circuit involves four switches A_1, A_2, B_1, and B_2, two supply connections, and two ground connections. The supply side switches, A_1 and B_1, are made of transmission-gates, a parallel connection of a 300 μm/0.16 μm PMOS and a 120 μm/0.16 μm NMOS device. The ground-side switches B_2 and A_2 are made of 120 μm/0.16 μm NMOS

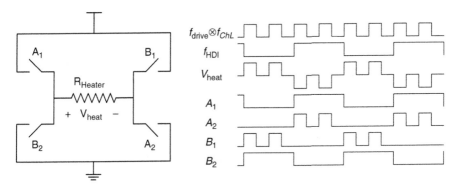

Fig. 6.20 The on-chip switching bridge driving the ETF heater, including the HDI scheme and the timing diagram of the switch drive signals

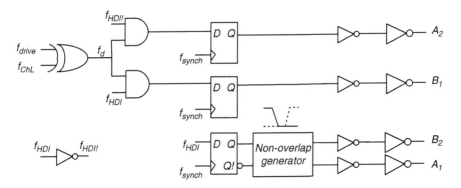

Fig. 6.21 The logic producing the heater driver switches

devices. The switches have a nominal on resistance of less than 7 Ω. The signals driving the switches are derived from f_{drive}, f_{HDI} and f_{ChL}. These are the ETF driving signal, the HDI drive, and the low-frequency chopping signal, respectively. The application of f_{ChL} to f_{drive} is simply done by means of an XOR gate (see Fig. 6.21 showing the digital circuitry producing the switch drive signals). For simplicity, the effect of f_{ChL} polarity change influencing the timing signals has not been shown in Fig. 6.20. The logic equations applicable to the switch drive signals are:

$$f_d = f_{drive} \otimes f_{ChL}$$
$$A_1 = \bar{f}_{HDI}$$
$$A_2 = f_d \ \& \ \bar{f}_{HDI}$$
$$B_1 = f_d \ \& \ f_{HDI}$$
$$B_2 = f_{HDI}. \tag{6.7}$$

Table 6.2 The expected operating points as well as expected accuracy and resolution considerations for the scaled electrothermal frequency reference

Block	Operating frequency	Expected inaccuracy	Expected resolution	Power	Noise or jitter
Scaled TD frequency reference	16 MHz	±0.1%	–	2 mW = 1 mW (ETF) + 1 mW (circuit)	50 ps (jitter)
ETF	1 MHz	0.1 degrees in phase	–	1 mW	12 nV/√Hz (output noise)
PDΔΣM	$f_{drive} = 1$ MHz $f_s = 41$ kHz $f_{NBW} = 10$ Hz	5 milli-degrees in phase	>13 bits	0.3 mW	10 nV/√Hz (input noise)
DCO	16 MHz	±10%	0.02%	0.2 mW	20 ps (jitter)
TS	2 Hz (conversion rate)	0.1°C	10 mK	0.2 mW	–

The switch driver signals are passed through large inverters to be able to drive the gate capacitances of the main bridge switches. The switch driver signals for A_1 and B_2, which negate each other, are passed through a non-overlapping signal generator to minimize the effect of shoot-through currents. The other signals (A_2 and B_1) are non-overlapping by logic (see Fig. 6.21), which means that in their case no shoot-through will happen.

The synchronization of the switch-drive signals has been done via a signal $f_{synch} = f_{DCO} = 16$ MHz (a copy of the DCO output signal). This signal then drives the synchronization D flip-flop between the phase DAC of the PDΔΣM and its synchronous demodulator (see Fig. 6.9a). Process, temperature and mismatch simulations performed on the spread of the delay between the heater current and the synchronous demodulator drive signal shows a worst-case delay of 15 ps, which is well below the 100 ps error budget.

A summary of the operating conditions as well as the expected accuracy and resolution of the major building blocks of the scaled electrothermal frequency reference is provided in Table 6.2.

6.6   Experimental Results

The scaled electrothermal frequency reference (Fig. 6.22) was fabricated in a baseline 0.16 μm CMOS process. The chip has an active area of 0.5 mm², which is 12× smaller than previous work. It dissipates 2.1 mW from a 1.8 V supply (1 mW in the ETF), which is 3.7× less than the previous design. 24 devices from one batch were built in ceramic DIL packages and characterized in a temperature-controlled oven.

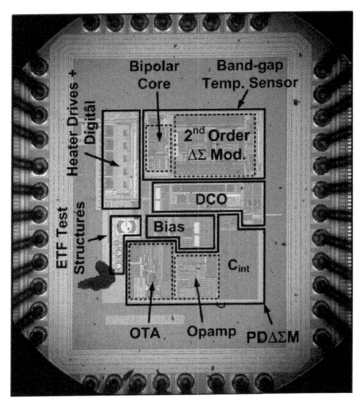

Fig. 6.22 Photomicrograph of scaled electrothermal frequency reference in 0.16 μm CMOS

Initially, a batch-calibration of 12 randomly selected devices was performed to determine the settings of the DCO's coarse DAC and to extract a 5th order compensation polynomial that will then be applied to the whole batch. This polynomial maps the digital temperature output μ of TS (Fig. 6.23b) to the 15-bit digital input Y of the D$\Delta\Sigma$M that, in turn, generates the $\phi_{ref}(T)$ (Fig. 6.23a) that maintains an output frequency of 16 MHz over the military temperature range: $-55°C$ to $125°C$. The measured device-to-device spread in ϕ_{ref} with respect to the FLL's master curve (Fig. 6.23a) represents a phase error of $\pm 0.15°$ (Fig. 6.24a). The measured device-to-device spread in μ with respect to the master curve of TS (Fig. 6.23b) represents an inaccuracy of about $\pm 0.25°C$ (Fig. 6.24b). The TS achieves 5mK (rms) resolution (at 1 conversion/s) and thus makes a negligible contribution (tens of ppm) to the frequency jitter.

After batch-calibration, the compensation polynomial (Fig. 6.25) was applied to 24 devices. Without trimming, the electrothermal frequency reference achieves an

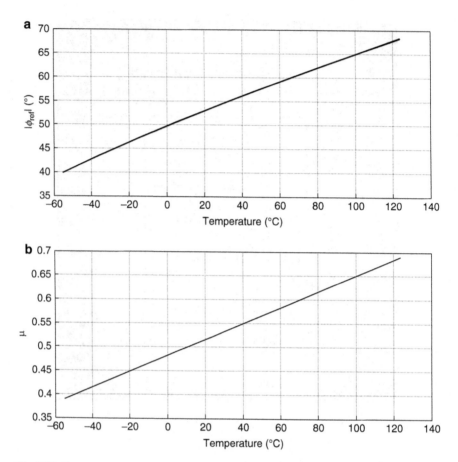

Fig. 6.23 Over temperature calibration of the scaled electrothermal frequency reference (**a**) ϕ_{ref} (T) that tunes f_{DCO} to 16 MHz; (**b**) The digital output of TS, µ

absolute frequency error of ±0.4% over the military temperature range (Fig. 6.26a). As in previous work, the PTAT spread of the TS can be trimmed together with the phase spread of ETF (see Chap. 5). This was done by using the PTAT trim knob implemented in the TS as a single trimming knob for the complete system. After a single room-temperature trim, the inaccuracy of the frequency reference drops to ±0.1% ($\sigma = \pm0.06\%$) (Fig. 6.26b), confirming that the accuracy of the previous design could be maintained. The residual temperature coefficient is low enough (< 12 ppm/°C via box method) to eliminate the need for temperature stability during the trimming process. As expected, the accuracy of the reference is limited by the inaccuracy of the TS (Fig. 6.27), which is ±0.5°C (3σ) and ±0.2°C (3σ) before and after trimming, respectively (24 samples).

In Fig. 6.28, the jitter of the scaled electrothermal frequency reference is shown. At 45 ps (rms) cycle-to-cycle jitter it is more than 7× less than that of the previous

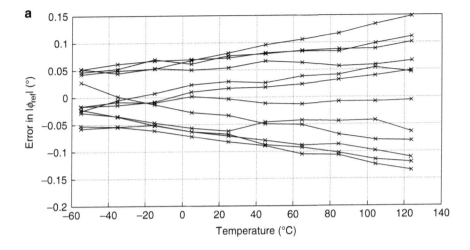

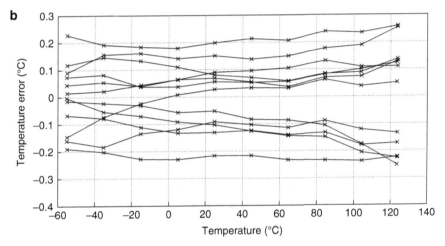

Fig. 6.24 Untrimmed over temperature errors at calibration phase (**a**) ϕ_{ref}; (**b**) TS temperature sensing error

design [6]. It is mainly limited by the 40 ps (rms) jitter of the free running DCO, which is larger than the 20 ps predicted by simulations. The excess jitter may be caused by supply noise or by errors in the $1/f$ noise corner models of the process.

In Table 6.3, the performance of the electrothermal reference is compared with that of other all-CMOS oscillators [6, 19–21, (see [34], Chap. 2)]. It may be seen that the scaled electrothermal frequency reference is more accurate than state-of-the-art RC oscillators and dissipates less power than an LC oscillator. It also

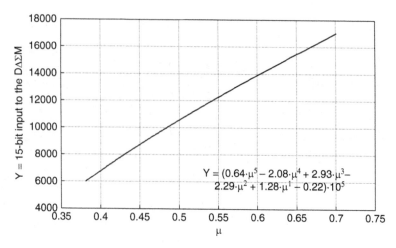

Fig. 6.25 Fifth-order polynomial that maps μ of TS to a 15-bit digital number Y that is input to DΔΣM

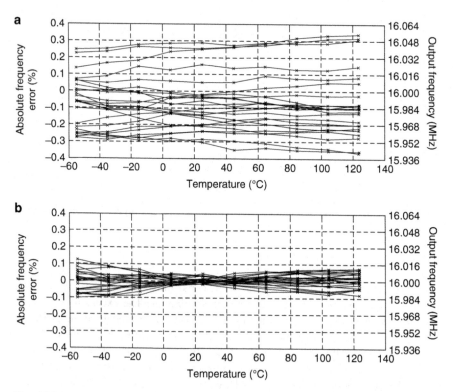

Fig. 6.26 Measured (24 devices) output frequency and its error from the target 16 MHz over military temperature range: (**a**) before trimming; (**b**) after a single room-temperature trim

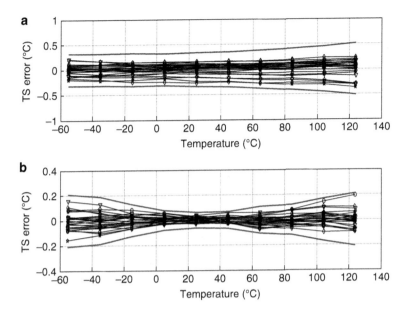

Fig. 6.27 The ultimate temperature sensing inaccuracy of TS (*bold lines* indicating the 3σ boundary of error): (**a**) untrimmed; (**b**) after a single room-temperature trim

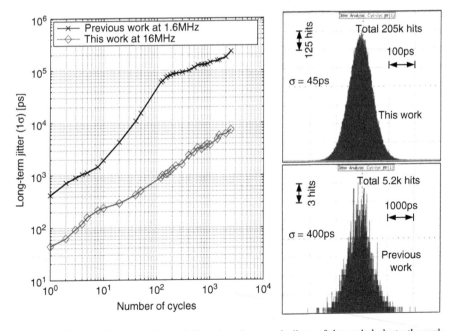

Fig. 6.28 Measured long-term jitter (*left*) and cycle-to-cycle jitter of the scaled electrothermal frequency reference (*top right*) in comparison with that of the previous design (*bottom right*)

Table 6.3 Performance summary and comparison to the current state-of-the-art

Reference	This work	[6]-TD	[20] and [21]-LC	[19]-RC
Frequency	16 MHz	1.6 MHz	24 MHz	14 MHz
Supply voltage (V)	1.8	5	1.8	1.8
Power consumption	2.1 mW	7.8 mW	<4 mW	45 μW
Technology	0.16 μm	0.7 μm	0.13 μm	0.18 μm
Area (mm²)	0.5	6.75	0.8	0.04
Temp. range (°C)	−55 ~ 125	−55 ~ 125	0~70	−40~125
Inaccuracy (ppm)	±1,000	±1,000	±50 to ±300	±1,900
Number of samples	24	16	Product	1
Period jitter (rms)	45 ps	312 ps	<1 ps	30 ps
Temp. coeff. (ppm/°C)	±11.2	±11.2	±1.4 to ±8.6	±23

maintains its accuracy over a wider temperature range with only a simple room temperature trim.

6.7   Conclusions

A scaled electrothermal (thermal-diffusivity-based) frequency reference with an output frequency of 16 MHz and 45 ps (rms) jitter has been implemented in a 0.16 μm CMOS process. The reference, occupying 0.5 mm², dissipates 2.1 mW from a 1.8 V supply and achieves an absolute inaccuracy of ±0.1% over the military temperature range (−55°C to 125°C). The improvements over the prior art prove that the performance of electrothermal frequency references benefits strongly from technology scaling.

References

1. Kent JP et al (2008) Microelectronics for the real world: 'Moore' versus 'More than Moore'. In: Proceedings of the IEEE custom integrated circuit conference (CICC), San Jose, CA, pp 395–402
2. Howard J et al (2010) A 48-Core IA-32 Processor in 45 nm CMOS using on-die message-passing and DVFS for performance and power scaling. IEEE J Solid-State Circ 46(1):173–183
3. Razavi B (2001) Design of analog CMOS integrated circuits. McGraw-Hill, New York
4. van Vroonhoven CPL et al (2010) A thermal-diffusivity-based temperature sensor with an untrimmed inaccuracy of ±0.2 °C (3σ) from −55 °C to 125 °C. In: IEEE ISSCC Dig. Tech. Papers, San Francisco, CA, pp 314–315
5. Kashmiri SM et al (2011) A scaled thermal-diffusivity-based frequency reference in 0.16 μm CMOS. In: IEEE 37th European solid-state circuits conference, ESSCIRC, Helsinki
6. Kashmiri SM et al (2010) A thermal-diffusivity-based frequency reference in standard CMOS with an absolute inaccuracy of ±0.1 % from −55 °C to 125 °C. IEEE J Solid-State Circ 45(12):2510–2520

7. Makinwa KAA, Snoeij MF (2006) A CMOS temperature-to-frequency converter with an inaccuracy of less than ±0.5 °C (3σ) from −40 °C to 105 °C. IEEE J Solid-State Circ 41(12):2992–2997
8. Makinwa KAA, Matova SP, Huijsing JH (2001) Thermopile design for a CMOS wind-sensor. In: National Dutch Sensor conference, The Netherlands, May 2001, pp 77–82
9. Huijsing JH (2011) Operational amplifiers, theory and design, 2nd edn. Springer, Dordrecht
10. Witte JF et al (2009) Dynamic offset compensated CMOS amplifiers. Springer, Dordrecht
11. Pertijs MAP, Huijsing JH (2006) Precision temperature sensors in CMOS technology. Springer, Dordrecht
12. Pertijs MAP et al (2008) Bitstream-controlled reference signal generation for a sigma-delta modulator. US Patent 7,391,351, June 2008
13. Schreier R, Temes GC (2005) Understanding delta-sigma data converters. Wiley, Hoboken
14. Zhang C, Makinwa KAA (2008) Interface electronics for a CMOS electrothermal frequency-locked-loop. IEEE J Solid-State Circ 43(7):1603–1608
15. Choe K et al (2009) A precision relaxation oscillator with a self-clocked offset-cancellation scheme for implantable biomedical SoCs. In: IEEE ISSCC Dig. Tech. Papers, San Francisco, CA, February 2009, pp 402–403
16. Souri K et al (2010) A CMOS temperature sensor with an energy-efficient zoom ADC and an Inaccuracy of ±0.25 °C (3σ) from −40 °C to 125 °C. In: ISSCC Dig. Tech. Papers, San Francisco, CA, February 2010, pp 310–311
17. Souri K et al (2010) A 0.12² mm2 7.4 μW micropower temperature sensor with an inaccuracy of ±0.2 °C (3σ) from −30 °C to 125 °C. In: Proceedings of the IEEE ESSCIRC 2010, Seville, Spain, pp 282–285
18. ADG719 data sheet. www.analog.com
19. McCorquodale MS et al (2010) A silicon die as a frequency source. In: Proceedings of the IEEE international frequency control symposium, June 2010, Newport Beach, California, pp 103–108
20. McCorquodale MS et al (2011) A history of the development of CMOS oscillators: the dark horse in frequency control. In: IEEE international frequency control symposium, San Francisco, CA, pp 437–442
21. Tokunaga Y et al (2010) An on-chip CMOS relaxation oscillator with voltage averaging feedback. IEEE J Solid-State Circ 45(6):1150–1158

Chapter 7
Conclusions and Outlook

This book has described the realization of frequency references based on the well-defined thermal diffusivity of IC-grade silicon. This chapter presents the main findings of the book. Furthermore, some suggestions for future work are presented.

7.1 Main Findings

- The well-defined thermal diffusivity of IC-grade silicon can be used as the basis for accurate on-chip frequency references in standard CMOS processes [1, 2].
- A digitally-assisted electrothermal frequency-locked loop (DAFLL) solves the integration challenges associated with earlier electrothermal frequency-locked loops (FLLs). This architecture facilitates the realization of electrothermal frequency references (Chap. 4).
- In a chopper amplifier with a gain-boosted output stage, the residual offset, emanating from the interaction between the output chopper's switching action, the cascode transistors' offset, and parasitic capacitances, can be suppressed by locating the chopper switches within the gain-boosting loops (Chap. 4).
- A DAFLL can be used as a test vehicle to measure the effective thermal diffusivity, D_{eff}, of substrate silicon. In a 0.7 μm CMOS process, the measured values at $-55°C$, $27°C$, and $125°C$ are 1.405, 0.755, and 0.495 cm^2/s, respectively. At room temperature, the value of D_{eff} is considerably lower than that of pure bulk silicon, which is probably due to the fact that the substrate consists of a lightly doped epitaxial layer on top of a heavily doped substrate, which has a higher thermal conductivity (Chap. 4).
- By measuring die temperature with a band-gap temperature sensor, a DAFLL can be temperature compensated. This resulted in the first integrated electrothermal frequency reference. Realized in a 0.7 μm CMOS process, this reference had an output frequency of 1.6 MHz and achieved an inaccuracy of $\pm 0.1\%$ from $-55°C$ to $125°C$ after a single room-temperature trim (Chap. 5).

S.M. Kashmiri and K.A.A. Makinwa, *Electrothermal Frequency References in Standard CMOS*, Analog Circuits and Signal Processing, DOI 10.1007/978-1-4614-6473-0_7, © Springer Science+Business Media New York 2013

- The performance of an electrothermal frequency reference is mainly determined by its electrothermal filter (ETF), which is embedded in a DAFLL. The ETF's accuracy is a function of lithographic inaccuracy, which improves with CMOS scaling. Scaling the ETF dimensions also enables trade-offs regarding the accuracy, power consumption, output frequency and jitter of an electrothermal frequency reference (Chap. 6).
- By scaling the ETF and adopting a more modern CMOS process, a 16 MHz scaled electrothermal frequency reference with 45 ps (rms) output jitter was implemented in a 0.16 μm CMOS process. Compared to the previous 0.7 μm realization, its output frequency is 10× higher, its power consumption is 3.7× lower, its area is 12× smaller and its output jitter is about 7× lower. The scaled reference maintains the same ±0.1% level of inaccuracy over the military temperature range, thus demonstrating that electrothermal frequency references benefit strongly from CMOS scaling.
- When compared to the state-of-the-art, [3–6], the accuracy of the proposed electrothermal frequency references is better than RC and ring oscillators, while their power consumption is lower and their operating temperature range is wider than LC oscillators.
- The inaccuracy of the proposed electrothermal frequency references is mainly limited by the accuracy of their temperature compensation schemes. State-of-the-art band-gap temperature sensors achieve inaccuracies in the order of 0.1°C [7], which translates into a frequency error of about 0.08% (Chaps. 5 and 6). This is because the temperature dependency of these frequency references is linked, via an ETF, to the strongly temperature dependent thermal diffusivity of silicon, D.

7.2 Future Work

The electrothermal frequency references proposed in this book could be further improved by means of future work at the system and circuit level. A summary of some possible improvements will be provided next.

Silicon-based electrothermal filters (ETFs) exhibit temperature drifts of about 0.3%/°C, which, together with the current accuracy of state-of-the-art temperature sensors, limits the frequency inaccuracy of electrothermal frequency references to about 0.1% [8–10]. Another material present in IC processes is silicon-dioxide, whose temperature dependence is about two orders of magnitude less than that of silicon [11, 12]. Perhaps the thermal path of an ETF could be partially or completely made in silicon-dioxide. This would reduce its temperature dependency and hence relax the requirements on the frequency reference's temperature compensation scheme.

Another way to improve the accuracy of the proposed electrothermal frequency references is by using a different trimming strategy. So far, this has been based on a cost-effective, single room-temperature trim. However, significant improvement can be expected if multi-point trimming is applied. It should be noted that the trimming resolution of the scaled frequency reference presented in Chap. 6 is limited to about 200 ppm by its 8-bit DCO. Thus an improved DCO is required if multi-point trimming is to be applied. As discussed in Chaps. 4 and 6, the DCO needs to have sufficient tuning range ($\pm 20\%$) to compensate for process spread and supply and temperature variations. In order to achieve both wide tuning range and fine resolution, a sigma-delta DAC can be combined with an oscillator to make a sigma-delta DCO [13].

Another way of improving the performance of the proposed electrothermal frequency references is to reduce their output jitter. This is mainly due to the ETF's thermal noise, which could be further reduced and its output signal increased by reducing its dimensions. This can be easily done by realizing the frequency reference in a CMOS process with even smaller feature sizes. Furthermore, a system-level modification could be applied to the DAFLL. The DCO jitter is currently suppressed by a first-order high-pass filter (see Chap. 3). Using a second-order filter would result in better jitter suppression. Another limiting factor is the jitter of the DCO (Chap. 6), which was based on a relaxation oscillator. Since a DAFLL can, in principle, utilize any type of DCO, lower jitter can be achieved by using oscillator types with lower inherent jitter, such as ring or LC oscillators.

Ultimately, the application of the above-mentioned improvements into the design of the future standard CMOS electrothermal (thermal-diffusivity-based) frequency references, should allow them to achieve inaccuracies in the order of 50 ppm, with lower power and jitter levels as well as over wider temperature ranges compared to the LC oscillators.

References

1. Kashmiri SM et al (2010) A thermal-diffusivity-based frequency reference in standard CMOS with an absolute inaccuracy of ± 0.1 % from -55 °C to 125 °C. IEEE J Solid-State Circ 45(12):2510–2520
2. Kashmiri SM et al (2012) A scaled thermal-diffusivity-based frequency reference in 0.16 μm CMOS. Invited to the special issue of IEEE J Solid-State Circ 47(7):1535–1545
3. SiTime's product selector sheet. Available online at: http://www.sitime.com/support/product-selector
4. De Smedt V et al (2009) A 66 μW 86 ppm/ °C fully-integrated 6 MHz wienbridge oscillator with a 172 dB phase noise FOM. IEEE J Solid-State Circ 44(7):1990–2001
5. McCorquodale MS et al (2011) A history of the development of CMOS oscillators: the dark horse in frequency control. In: IEEE international frequency control symposium, San Francisco, CA, pp 437–442
6. Sundaresan K et al (2006) Process and temperature compensation in a 7-MHz CMOS clock oscillator. IEEE J Solid-State Circ 41(2):433–442
7. Pertijs MAP, Huijsing JH (2006) Precision temperature sensors in CMOS technology. Springer, Dordrecht
8. Kashmiri SM et al (2009) A temperature-to-digital converter based on an optimized electro-thermal filter. IEEE J Solid-State Circ 44(7):2026–2035

9. Makinwa KAA, Snoeij MF (2006) A CMOS temperature-to-frequency converter with an inaccuracy of less than ±0.5 °C (3σ) from −40 °C to 105 °C. IEEE J Solid-State Circ 41(12):2992–2997

10. Kashmiri SM, Makinwa KAA (2009) A digitally-assisted electrothermal frequency-locked loop. In: Proceedings of the 35th ESSCIRC, Athens, Greece, pp 296–299

11. Vermeersch B (2009) Thermal AC modelling, simulation and experimental analysis of microelectronic structures including nanoscale and high-speed effects. Ph.D. dissertation, University of Gent

12. Ju YS, Goodson KE (1999) Process-dependent thermal transport properties of silicon-dioxide films deposited using low-pressure chemical vapor deposition. J Appl Phys 85(10):7130–7134

13. Khalil W et al (2011) A 700-μA 405-MHz all-digital fractional-N frequency-locked loop for ISM band applications. IEEE Trans Microw Theory Tech 59(5):1319–1326

Appendix
Time-Domain Modelling of an Electrothermal DAFLL

This appendix provides a time-domain modeling method for the digitally-assisted electrothermal frequency-locked loop (DAFLL). A set of basic Matlab codes have been developed, which can be used in further simulations of various time-domain effects in the behavior of a DAFLL. The electrothermal filter (ETF) of the loop realizes part of its signal chain in the thermal-domain. It will be shown how the analogy between the thermal and electrical domains can be used to develop a network of resistors and capacitors that reproduce the ETF characteristic. This network can be embedded in the time-domain model of the loop, which is in principle an analog mixed-mode system, with continuous-time and discrete-time functions. These can be modeled with the behavioral simulation techniques used for data converters.

A.1 Time-Domain Simulation of an ETF

Chapter 3 provided an analytical modeling of the ETF based on its thermal impedance (see Sects. 3.3 and 3.4). This model mainly describes the frequency-domain characteristics of an ETF by means of numerical calculations, which are rather difficult to use in time-domain simulations. It should be noted that the thermal-impedance model has been extensively used in determining the various sensitivity functions that relate the DAFLL performance to various error sources.

A simpler method for the time-domain modeling of an ETF is to match the frequency-domain response of a low-pass filter, e.g. a network of R and C elements (electrical-domain resistors and capacitors), to that of the ETF. This method was previously used in the development of a *wind sensor* that incorporated electrothermal filters. The step response measured for the wind sensor was fitted to that of a Foster network of 20 parallel RC segments (see Fig. A.1). Here, the wind sensor model has been adopted and its RC values have been modified to map its phase, amplitude

S.M. Kashmiri and K.A.A. Makinwa, *Electrothermal Frequency*
References in Standard CMOS, Analog Circuits and Signal Processing,
DOI 10.1007/978-1-4614-6473-0, © Springer Science+Business Media New York 2013

Fig. A.1 The Foster
equivalent RC network of the
ETF that is used in the time-
domain simulations

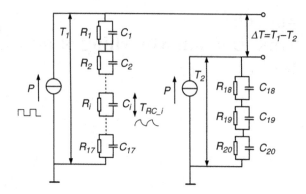

and step response to that of the ETF, as predicted by the thermal-impedance model
(see Fig. 3.16). The resulting R and C values are reported in Table 3.2.

The Foster network shown in Fig. A.1 is excited by the square-wave current
sources with amplitude P (ETF heater power). This current (power dissipation)
results in an AC voltage (AC temperature variations) at the output of the network
(thermopile output). This output is denoted by ΔT, and is the difference between the
voltages (temperatures) T_1 and T_2, which represent the *hot* and *cold* junction
temperatures of the thermopile, respectively:

$$\Delta T = T_1 - T_2. \tag{A.1}$$

Using super positioning, ΔT can be written as a function of each RC segment's
voltage (temperature):

$$\Delta T = \sum_{i=1}^{17} T_{RC\_i} - \sum_{j=18}^{20} T_{RC\_j}. \tag{A.2}$$

The goal is to calculate the ΔT (thermopile output) in the time-domain using the
P, R_i and C_i values, and the heater excitation frequency. Simulations were carried
out in Matlab (A.3), where numbers are mainly stored and processed in vectors with
an integer number of elements. This means that the time-domain simulations can be
done at a limited number of instances, which introduces quantization in time. As a
result, a minimum time-step of Δt can be defined as the time space between every
two calculation instances of ETF output.

In order to calculate T_{RC_i} for each segment in (A.2), the step response of a single
RC element (shown in Fig. A.2) is considered. The voltage (temperature) across
this parallel RC segment, denoted by $T_{RC}(t)$, when excited by a step current source
$i(t) = P \cdot u(t)$ is:

$$T_{RC}(t) = P \cdot R \cdot (1 - e^{-t/\tau}). \tag{A.3}$$

where $\tau = R \cdot C$.

Fig. A.2 Discrete-time
values of an RC segment step
response

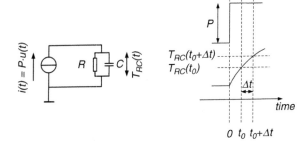

As shown in Fig. A.2, $T_{RC}(t)$ can be calculated at discrete time instances, each Δt seconds apart. The goal is to derive a *generic* equation for $T_{RC}(t)$ that can be plugged into a *generic* Matlab code for calculation of ETF output signal at discrete time instances. To do so, parameter t can be expanded into a discrete equation:

$$t = t_0 + n \cdot \Delta t. \tag{A.4}$$

with t_0 as an initial moment of time and n as an integer. In the next step, $T_{RC}(t)$ should be calculated at two consecutive instances: t_0 and $t_0 + \Delta t$ (see Fig. A.2). This allows $T_{RC}(t_0 + \Delta t)$ to be derived as a function of $T_{RC}(t_0)$, which enables the compilation of a *generic* discrete-time equation for $T_{RC}(n)$. Using (A.3):

$$T_{RC}(t_0 + \Delta t) = P \cdot R \cdot \left(1 - e^{\frac{-(t_0 + \Delta t)}{\tau}}\right). \tag{A.5}$$

which can be modified by adding and subtracting an extra term:

$$T_{RC}(t_0 + \Delta t) = P \cdot R \cdot \left(1 - e^{\frac{-(t_0 + \Delta t)}{\tau}}\right) + P \cdot R \cdot e^{\frac{-\Delta t}{\tau}} - P \cdot R \cdot e^{\frac{-\Delta t}{\tau}}. \tag{A.6}$$

By re-arranging:

$$T_{RC}(t_0 + \Delta t) = P \cdot R \cdot \left(1 - e^{\frac{-t_0}{\tau}}\right) \cdot e^{\frac{-\Delta t}{\tau}} + P \cdot R \cdot \left(1 - e^{\frac{-\Delta t}{\tau}}\right). \tag{A.7}$$

The first term on the right hand side of (A.7) is equal to $T_{RC}(t_0)$ and thus:

$$T_{RC}(t_0 + \Delta t) = T_{RC}(t_0) \cdot e^{\frac{-\Delta t}{\tau}} + P \cdot R \cdot \left(1 - e^{\frac{-\Delta t}{\tau}}\right). \tag{A.8}$$

Using (A.4), (A.8) can be expanded into an equation for discrete samples $n = 1$, $2, 3, \ldots$ such that:

$$T_{RC}(n) = T_{RC}(n - 1) \cdot exv + P \cdot R \cdot (1 - exv). \tag{A.9}$$

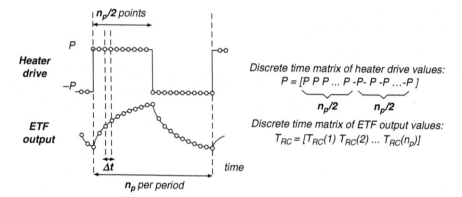

Fig. A.3 An illustration of the ETF continuous and discrete-time input and output waveforms, and matrixes that could be programmed in Matlab to hold the discrete-time values

where

$$exv = e^{\frac{-\Delta t}{\tau}}. \tag{A.10}$$

Within a complete period of the ETF heater drive defined by $T_{drive} = 1/f_{drive}$, a total number of n_p discrete time instances can be considered (see Fig. A.3). With this assumption, the value of Δt can be calculated as:

$$\Delta t = \frac{T_{drive}}{n_p}. \tag{A.11}$$

Furthermore, the heater drive signal $P(n)$ can be written as:

$$P(n) = \begin{cases} P & for & 1 \leq n < \dfrac{n_p}{2} \\ -P & for & \dfrac{n_p}{2} \leq n < n_p \end{cases}. \tag{A.12}$$

where $+P$ and $-P$ result in a differential thermopile output signal (see Figs. 3.7 and 3.8 for fully differential thermopile structures).

Using (A.2), (A.9), and (A.10), a *generic* equation can be developed, which calculates the ETF output signal within a period of the heater drive:

$$\underbrace{\Delta T(n)}_{n=1,2,..,n_p} = \sum_{i=1}^{17} T_{RC\_i}(n-1) \cdot exv_i + P(n) \cdot R_i \cdot (1 - exv_i)$$

$$- \sum_{j=18}^{20} T_{RC\_j}(n-1) \cdot exv_j + P(n) \cdot R_j \cdot (1 - exv_j). \tag{A.13}$$

where

$$exv_k = \frac{-\left(\frac{T_{drive}}{n_p}\right)}{R_k \cdot C_k}.$$ (A.14)

The RC values reported in Table 3.2 for the ETF model can be loaded into matrixes to be used in a *generic* Matlab code that calculates the ETF output for a period of the heater drive:

$$R = [R_1 \ R_2 \ R_3 \ \dots \ R_{17} \ R_{18} \ R_{19} \ R_{20}]$$
$$C = [C_1 \ C_2 \ C_3 \ \dots \ C_{17} \ C_{18} \ C_{19} \ C_{20}].$$ (A.15)

The result is the following Matlab code that can be extended to any number of periods by extending the vectors that define the heater drive input:

```
tau = R.*C;                  % time constant vector
np = 100;                    % number of points per heater drive period
fdrive = 100e3;              % heater drive frequency = 100kHz in this
                             case
Tdrive = 1/fdrive;           % heater drive period
delta_t = Tdrive/np;         % time step
power = 2.5e-3;              % heater power = 2.5mW in this case
P(1:np/2) = power;           % first half period of heater drive
P(np/2+1:np) = -power;       % second half period of heater drive
s_tp = 24*0.5e-3;            % number of thermocouples times Seebeck
                             coefficient of one
pol(1:17)=1;                 % summation polarity for hot junction side
                             of network
pol(18:20)=-1;               % summation polarity for cold junction
                             side of network
exv = exp(-delta_t./tau);    % realization of exv_k in equation A.14
dT((1:20),1) = 0;
for i=2:length(P);           % This loop runs per calculation point of
                             the ETF output
  dT(:,i) = dT(:,(i-1)).*exv +
    P(i)*R.*(1-exv);         % realization of sigma functions in A.13
  V_ETF(i) = pol*
    dT(:,i)*s_tp;            % translation to voltage and summation
end;                         % At the end, the ETF output signal is
                             stored in V_ETF matrix
```

Simulation results for the ETF output signal based on this code, and within a few periods of the heater drive at 100 kHz, and at two levels of the heater power are shown in Fig. 3.35. This code will be embedded into another code that builds the complete time-domain model of the DAFLL. This will be discussed in the following section.

A.2 Time-Domain Simulation of the DAFLL

The DAFLL system architecture was described in Chap. 4, as depicted by the block-diagram of the loop shown in Fig. 4.25. This loop includes a phase-domain $\Delta\Sigma$ modulator (PD$\Delta\Sigma$M), a 12-bit digitally-controlled oscillator (DCO), and a digital integrator. The block-diagram that formed the basis on which the Matlab model of the loop is developed and shown in Fig. A.4.

The output of the RC network representing ETF is fed to the PD$\Delta\Sigma$M, where it is multiplied by the output of a phase DAC. The result is integrated by the continuous-time integrator of the modulator, V_integ_PDSD, which drives the quantizer. The PD$\Delta\Sigma$M, the DAFLL's digital integrator, and the DAC driving the VCO, are sampled at frequency f_s. The DAC has a total number of bits $= N_{DAC}$ and a reference voltage $= V_{ref}$. The VCO has a voltage-to-frequency gain $= K_{VCO}$. The output frequency of the VCO updates a parameter in the loop: f_{VCO}, from which f_{drive} of ETF and f_s, are extracted.

Matlab is a discrete-time environment in which the values of various signals in the loop can be processed and stored in vectors. The loop is simulated for a limited number of sampling periods $= N_{sample}$. Each sampling period includes an integer number of ETF heater drive periods $= 2 \times N_{fs}$, where N_{fs} represents the number of half-periods. Per half-period of ETF drive signal, there are a limited number of time instances $= N_{points}$. The continuous analog signals within the loop, e.g. the ETF output, V_ETF, and V_integ_PDSD, are only calculated within these time instances. Therefore, the time interval among each two calculation points, denoted by Δt, is the smallest time interval in the whole calculations. In every simulation run of DAFLL, the total number of calculation points is equal to:

$$N_{total} = \underbrace{N_{sample}}_{\substack{\text{Total number of samples}}} \times \underbrace{2 \times N_{fs}}_{\substack{\text{Total number of ETF drive periods} \\ \text{per loop sample}}} \times \underbrace{N_{points}}_{\substack{\text{Total number of time instances in} \\ \text{one half period of ETF drive}}}$$

$$(A.16)$$

The minimum time interval Δt of the loop is updated every time the DCO output frequency is updated. The value of Δt remains the same (for the calculation of the intermediate points) until the next sampling moment of the loop, when the DCO frequency is updated. Considering a division by 16 between f_{VCO} and f_{drive} (see Fig. A.4), the value of Δt can be calculated from:

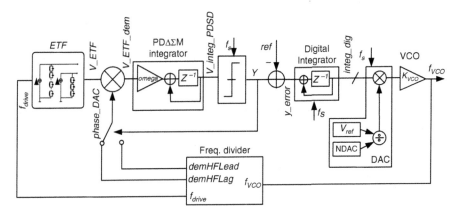

Fig. A.4 Block-diagram of DAFLL used for Matlab simulations

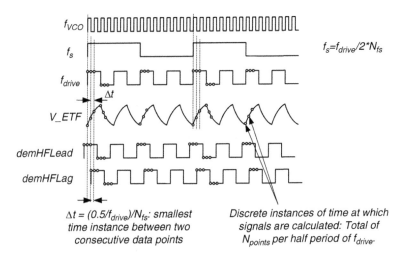

Fig. A.5 Some of discrete-time instances within DAFLL continuous-time signals, and the time interval Δt between them

$$\Delta t = 1/(16 \cdot f_{VCO} \cdot 2 \cdot N_{points}). \qquad (A.17)$$

All the periodic signals within the loop can be defined in matrixes with an integer number of elements. For instance: f_{drive}, the signal that drives the ETF heater, can be defined by a vector with an N_{total} number of elements [see (A.16)]. In this vector, the value of elements toggle periodically between $+P$ and $-P$ (P is the heater power) at every N_{points} element. The phase DAC reference signals can also be defined in the same way. These are equivalent to the f_{drive} vector, with the difference being that their elements toggle between $+1$ and -1. The lead and lag phase shifts of the phase DAC references, in reference to f_{drive}, can be simply implemented by shifting the vector elements forwards and backwards (see Fig. A.5) by a total number of:

$$N_{shift} = 2 \cdot N_{points} \cdot \frac{phase\_ref}{360°}. \qquad (A.18)$$

where *phase_ref* can for instance be equal to $\pm 45°$.

Once the signals in the DAFLL system block diagram are pre-defined, an algorithm (see Fig. A.6) can be defined that runs for the total number of samples and calculates the intermediate points for the internal signals. This algorithm involves three loops. The *most inner loop* runs per half-period of the ETF heater drive signal. This is a *for loop* in Matlab that runs for N_{points}. Within this loop, the ETF output signal $V_{ETF}(n)$, and the result of its multiplication with the phase DAC output, $V_{ETF\_dem}(n)$, are calculated. Furthermore, this loop calculates the PD$\Delta\Sigma$M's integrator output, V_integ_PDSD. Consequently, the integrator's continuous-time output signal is calculated every time one of the intermediate ETF output signal points is calculated. This reproduces the continuous-time integrator's signal in an over-sampled fashion in regard to the modulator's sampling frequency.

An ideal continuous-time integrator has a frequency-domain transfer:

$$H(s) = \frac{\omega_0}{s}. \qquad (A.19)$$

in which ω_0 is its unity-gain frequency (Fig. A.7a), i.e. the frequency at which the amplitude response of the integrator reaches the 0 dB gain point. In the time-domain, the output of such integrator will be a ramp in response to a unit step input (Fig. A.7b). After Δt seconds the integrator output growth (from zero initial condition) is:

$$\Delta V_{out}(\Delta t) = \omega_0 \cdot \Delta t. \qquad (A.20)$$

In the discrete time simulation environment of Matlab, a calculation of the continuous-time integrator output at the *ith* discrete sample can be written as:

$$V_{out}(i) = V_{out}(i-1) + v_{in}(i) \cdot \omega_0 \cdot \Delta t. \qquad (A.21)$$

For the DAFLL, Δt can be calculated from (A.17).

The *intermediate loop* of the DAFLL algorithm (see Fig. A.6) runs for the f_{drive} half periods within one sampling period of the loop. Therefore, the loop runs for $2 \cdot N_{fs}$ times between every two samples. The *most outer loop* runs for the total number of samples N_{sample}. Within this loop, the PD$\Delta\Sigma$M quantizer and the phase DAC outputs are updated. Also within the same loop, the bitstream of the modulator, Y (see Fig. A.4), is compared with the DAFLL phase reference, *ref*. This is a consecutive set of one's and zero's at a rate of $f_s/2$ and represents a 90° phase shift in the ETF (for *phase_ref* values of ± 45). The result of this comparison is integrated by the digital integrator, whose output then updates the DAC input

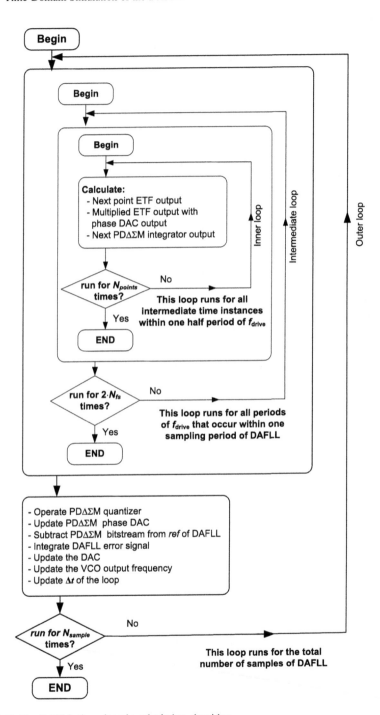

Fig. A.6 The DAFLL time-domain calculation algorithm

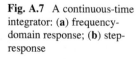

Fig. A.7 A continuous-time integrator: (**a**) frequency-domain response; (**b**) step-response

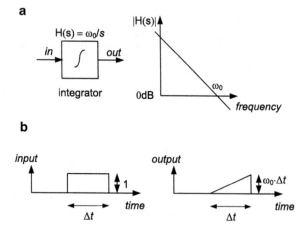

every time the outer loop runs. This then changes the VCO output frequency, f_{VCO}, which updates f_{drive} and Δt eventually.

Based on the proposed algorithm a Matlab code can be devised which forms the basis for the time-domain simulations of the DAFLL shown in Fig. A.4. This code can be further expanded for the addition of non-idealities or the study of loop dynamics in the time domain. A few simulation results, such as the loop startup and step response, as well as the output spectrum of the PD$\Delta\Sigma$M and digital integrator, were shown in Chap. 4 and in Figs. 4.26 and 4.27. The generic Matlab code that implements the proposed algorithm is as follows:

```
% Initial definitions
s_tp = 24*0.5e-3;            % Thermopile sensitivity
kvco = 0.9e6;               % VCO sensitivity [Hz/V]
Nsample = 8192;             % Number of samples of the loop
Nfs = 2;                    % Number of ETF drive periods per sample
Nperiods = 2*Nfs*Nsample;   % Number of half periods of ETF drive to be
                              simulated
Npoints = 24;               % Number of intermediate points in every
                              half period of fdrive
f = 100e3;T = 1/f;          % Initial condition for ETF drive frequency
fvco = f*ones(1,Nsample);   % Define a vector that holds fVCO values
vco_in = zeros(1,Nsample);  % Define a vector holding VCO input signal
fs = f/Nfs;                 %Sampling frequency of DAFLL
fT = 300e3;                 % 0-dB frequency of the continuous-time
                              integrator [Hz]
NDAC=12;Vref=4;             % DAC definition
```

```
demHFLead = zeros (1, Nperiods*Npoints);   % phase DAC's Lead feedback
                                             signal
demHFLag = zeros (1, Nperiods*Npoints);    % phase DAC's Lag feedback
                                             signal
V_integ_PDSD =
  zeros(1, Nperiods*Npoints);               % PDSD's integrator signal
y = ones (1, Nsample);                      % PDSD's bitstream

% DAFLL reference bit stream (90 degrees phase), and error signal of
the loop
yerror = zeros(1,Nsample);
diginteginit = 256;       % The initial condition at which the digi-
                            tal integrator starts
integ_dig = diginteginit*
  ones(1,Nsample);        % error signal pre-allocation
ref = zeros(1,Nsample);   % DAFLL reference signal
ref(3:2:Nsample)=1;       % DAFLL reference signal

% Definition of the ETF drive square-wave
range  =  (1:2*Npoints: Nperiods*Npoints);  %  initialization  of
                                                 vectors
fdrive = zeros (1, Nperiods*Npoints);     % initialization of vectors
fdrive(range) = 1;                        % initialization of vectors
cycle = [ones(1,Npoints),
  zeros(1,Npoints)];                      % initialization of vectors
fdrive = filter
  (cycle,1,fdrive);                       % initialization of vectors
fdrive = 2*fdrive - 1;
demHFLead = circshift (fdrive, [1 -Npoints/4]);   % Definition of the
                                                    Lead feedback
demHFLag = circshift (fdrive, [1 Npoints/4]);     % Definition of the
                                                    lag feedback
phase_DAC = demHFLead;                            % Initial phase DAC
                                                    value
P = 25/(2*1.2e3);         % The heater drive power [W]
R = etf_model(:,2);       % The vector including ETF's Foster net-
                            work resistances
C = etf_model(:,3);       % The vector including ETF's Foster net-
                            work capacitances
pol(1:17)=1;              % polarity of summation for hot junction
                            side of network
pol(18:20)=-1;           % polarity of summation for cold junction
                            side of network
tau = R.*C;              % Time constant vector
V_ETF = zeros (1,Nperiods*Npoints);         % pre allocating thermopile
                                              output signal
V_ETF_dem= zeros (1,Nperiods*Npoints);      % pre allocating demodulated
                                              ETF output

ix = 1;
% Assignment of startup values so that the loop can start working
delta_t = T/(2*Npoints);
```

```
exv = exp(-delta_t./tau);
dT((1:20),1) = 0;
% An initial condition for the continuous time integrator's cut off
frequency
omega = 2*pi*fT/fs/(2*Npoints*Nfs);
% The DAFLL loop simulation
% The outer loop runs per loop sample
for sdx = 2:Nsample
% The mid loop runs per half period of fdrive in one sampling period
of sigma-delta
  for jx = 2:2*Nfs+1
    % The inner loop runs per intermediate calculation steps
    for px = 1:Npoints
    % Calculating the ETF output value for the current intermediate
    point
    ix = ix + 1;
    ix = min(Npoints*2*Nfs*Nsample,ix);
    dT(:,ix) = dT(:,(ix-1)).*exv + P*fdrive(ix)*R.*(1-exv);
    V_ETF(ix) = pol*dT(:,ix)*s_tp;
    % multiply ETF output with phase DAC feedback phase
    V_ETF_dem(ix) = V_ETF(ix)*phase_DAC(ix);
    % phase-domain delta sigma's integrator value evaluation
    V_integ_PDSD(ix) = V_integ_PDSD(ix-1) + omega*V_ETF_dem(ix);
    end
  end
  % Quantizer
  y(sdx) = sign(V_integ_PDSD(ix));
  % Phase DAC
  if (y(sdx)==1)
    phase_DAC = demHFLead    else
    phase_DAC = demHFLag   end;
% DAFLL phase summation node: subtraction, error signal generation
yerror(sdx) = ref(sdx)- (((y(sdx)*-1)+1)/2);
%subtract PDSD bitstream & phase ref
integ_dig(sdx) = integ_dig(sdx-1) + yerror(sdx-1);   %digital integrator

% Update the DCO DAC value
vco_in(sdx) = 4*(integ_dig(sdx))*(Vref/2^(NDAC-1));
% Update the VCO frequency
fvco(sdx) = max(kvco*vco_in(sdx),100e3);
% Feedback the VCO output to the ETF, i.e. close the DAFLL loop
feedback
f = fvco(sdx)/16;
T = 1/f;
fs = f/Nfs;
```

```
delta_t = T/(2*Npoints); % update the time stamp of the loop based
on latest fVCO
exv = exp(-delta_t./tau);
% Update the PDSD's continus time integrator cut off frequency
omega1 = 2*pi*fT/fs/(2*Npoints*Nfs);
end
```

Summary

This book investigates the concept of generating accurate on-chip frequencies based on the thermal properties of silicon. The work presented here shows that electrothermal (thermal-diffusivity-based) frequency references are feasible in standard CMOS processes. Furthermore, the possibility of scaling such references into the more modern CMOS processes is studied. The improvements achieved in the performance of the scaled reference show that such frequency references strongly benefit from process scaling.

A frequency reference is an indispensable part of most electronic systems. In recent years, a lot of effort has been devoted to the replacement of quartz crystal oscillators with integrated solutions. The main motivations have been the reduction in size and cost as well as the increase in reliability of electronic circuits. Chapter 2 provides an overview of the recent developments in the implementation of silicon-based frequency references. In this survey, various types of oscillators including MEMS, RC, LC, ring, and electron-mobility-based are studied. The MEMS and LC oscillators have been commercialized, so far. These references achieve stabilities of < 1 ppm and 50 ppm, respectively, enabling them to compete with the quartz crystal oscillators.

Apart from some earlier thermal oscillators, the thermal properties of silicon have received much less attention in generation of on-chip frequencies. These concepts are discussed in Chap. 3. In principle, the heat generated in a silicon substrate diffuses at a defined rate which is determined by the thermal diffusivity of silicon, D. This forms the basis for the thermal delay lines, around which the thermal oscillators are made. A thermal delay line embeds a heater at a distance s from a temperature sensor within a silicon substrate. Its delay is then governed by D and s. The former is a stable property of IC-grade silicon, while the latter is defined by lithography. The early thermal oscillators had large levels of output jitter and high levels of power consumption. These were mainly due to heat losses in the substrate, which resulted in very small signals at the output of their temperature sensors.

Chapter 3 further describes how a thermal delay line can be extended into a more complex structure called an electrothermal filter (ETF). An ETF realizes a heater

S.M. Kashmiri and K.A.A. Makinwa, *Electrothermal Frequency*
References in Standard CMOS, Analog Circuits and Signal Processing,
DOI 10.1007/978-1-4614-6473-0, © Springer Science+Business Media New York 2013

(e.g. a resistor) at a close distance s (less than 25 μm) from a relative temperature sensor (e.g. a thermopile), and thus can be realized in standard CMOS processes. This structure behaves like a low-pass filter, whose well-defined phase response is determined by D and its geometry. Measurements on ETFs have shown tolerances in the order of 0.1%. Chapter 3 describes how embedding an ETF into a feedback loop with a narrow noise-bandwidth solves the unacceptable jitter performance of the previous thermal oscillators. Such loop is called an electrothermal frequency-locked loop (FLL). An FLL locks the output frequency of a VCO to the accurate phase shift of the ETF. As a result, this frequency is determined only by the ETF and thus not affected by the VCO's tolerances and drifts. Earlier implementations of electrothermal FLLs are investigated in Chap. 3. Since the output frequency of an FLL follows the same $T^{-1.8}$ temperature dependence of D, these FLLs were aimed as accurate temperature-to-frequency converters.

The temperature-dependent output frequency of FLLs showed promising levels of untrimmed device-to-device frequency spread, which was in the order of $\pm 0.25\%$ over the industrial temperature range. Chapter 3 concludes that such levels of inaccuracy imply that a temperature-compensated electrothermal FLL should form a suitable basis for the realization of accurate electrothermal frequency references.

There were challenges associated with the integration of early FLLs. These required large off-chip capacitors in order to implement their narrow noise-bandwidths, making their CMOS integration rather difficult. Therefore, Chap. 4 proposes a digitally-assisted electrothermal FLL (DAFLL), in which the narrow noise-bandwidth of the loop is implemented by a digital filter. A DAFLL incorporates a digital phase detector (a phase-domain $\Delta\Sigma$ modulator) that digitizes the phase shift of an ETF, a digital phase reference, a phase summation node, and a digitally-controlled oscillator (DCO). An implementation in a standard 0.7μm CMOS achieved an output frequency with an untrimmed device-to-device spread of $\pm 0.3\%$ (3σ) from $-55°C$ to $125°C$. This level of inaccuracy is comparable to the early FLLs. Furthermore, the output frequency of the loop followed the same $T^{-1.8}$ temperature dependence, associated with the thermal diffusivity of silicon. The DAFLL provided a solution for the integration challenge of electrothermal FLLs and therefore was adopted as the foundation of the electrothermal frequency references proposed in this book.

Chapter 5 describes how an electrothermal frequency reference can be realized by temperature-compensating a DAFLL. This was done by measuring the temperature of the die using an on-chip band-gap temperature sensor and injecting the digital temperature information into the loop. The result was the first fully-integrated electrothermal (thermal-diffusivity-based) frequency reference, realized in a standard 0.7 μm CMOS process. The reference has an output frequency of 1.6 MHz, dissipates 7.8 mW, and achieves an absolute inaccuracy of $\pm 0.1\%$ over the military temperature range ($-55°C$ to $125°C$) with a single room-temperature trim. Its worst-case temperature coefficient of ± 11.2 ppm/°C allows for trimming without temperature stabilization, which simplifies the trimming procedure and thus reduces trimming costs.

Chapter 6 describes the scaling of an electrothermal frequency reference into a more modern CMOS process. The performance of an electrothermal frequency reference is mainly determined by its ETF. The accuracy and phase-frequency characteristic of an ETF are functions of its geometry. Scaling the ETF enables trade-offs regarding the accuracy, power consumption, output frequency and jitter. By scaling the ETF and adopting a more modern CMOS technology, a scaled electrothermal frequency reference was developed in a standard 0.16 μm CMOS process. The ETF's critical dimension s (the distance between heater and temperature sensor) was scaled from 24 to 4.7 μm, which results in 5.5× more SNR and allows for a 10× higher frequency. The improved lithographic accuracy of the 0.16μm process (compared to the previous 0.7 μm process) maintains its accuracy. The resulting frequency reference has an output frequency of 16 MHz (10× higher compared to previous work), occupies an area of 0.5 mm^2 (12× smaller), dissipates 2.1 mW (3.7× less), and has an rms output jitter of 45 ps (7× less). Measurements on 24 devices show that the reference maintains the ±0.1% level of inaccuracy (from −55°C to 125°C) achieved by the previous work with a single room-temperature trim.

About the Authors

S. Mahdi Kashmiri was born on March 21st, 1980 in Tehran, Iran. He received his B.Sc. degree in electrical engineering from Tehran University, Tehran, Iran, in 2001, and his M.Sc. degree in microelectronics (*cum laude*) from Delft University of Technology, Delft, The Netherlands in 2006. In April 2012, he received his Ph.D. degree from the same university for his work on electrothermal frequency references in standard CMOS. From June 2001 to August 2004, he was a system design engineer at Parman Co. Tehran, Iran, where he worked on the development of a SDH fiber optics telecommunication system. From September 2005 to October 2006, he was an intern at the mixed-signal circuit and systems group of Philips Research Laboratories (currently NXP research), Eindhoven, The Netherlands, where he worked on a wide-band continuous-time sigma-delta modulator. Since October 2010 he has been with the precision systems group of Texas Instruments Incorporated, Delft, The Netherlands (formerly National Semiconductor Corporation). His research interests include the analog and mixed-signal integrated circuits, precision analog systems, and data converters.

Mr. Kashmiri received the 2009 Young Scientist Award of the European Solid-State Circuits Conference (ESSCIRC).

Kofi A.A. Makinwa received the B.Sc. and M.Sc. degrees from Obafemi Awolowo University, Nigeria in 1985 and 1988 respectively. In 1989, he received the M.E.E. degree from the Philips International Institute, The Netherlands and in 2004, the Ph.D. degree from Delft University of Technology, The Netherlands.

From 1989 to 1999, he was a Research Scientist with Philips Research Laboratories, Eindhoven, The Netherlands, where he worked on interactive displays and on front-ends for optical and magnetic recording systems. In 1999, he joined Delft University of Technology, where he is now an Antoni van Leuwenhoek Professor in the Faculty of Electrical Engineering, Computer Science and Mathematics. His main research interests are in the design of precision analog circuitry, sigma-delta modulators, smart sensors and sensor interfaces. This has resulted in 4 books, 18 patents and over 150 technical papers.

S.M. Kashmiri and K.A.A. Makinwa, *Electrothermal Frequency*
References in Standard CMOS, Analog Circuits and Signal Processing,
DOI 10.1007/978-1-4614-6473-0, © Springer Science+Business Media New York 2013

Kofi Makinwa is on the program committees of the European Solid-State Circuits Conference (ESSCIRC) and the Advances in Analog Circuit Design (AACD) workshop. He has also served on the program committee of the International Solid-State Circuits Conference (ISSCC), as a guest editor of the Journal of Solid-State Circuits (JSSC) and as a distinguished lecturer of the IEEE Solid-State Circuits Society (2008–2011). He is a co-recipient of several best paper awards: from the JSSC, ISSCC, Transducers and ESSCIRC, among others. In 2005, he received a Veni Award from the Netherlands Organization for Scientific Research and the Simon Stevin Gezel Award from the Dutch Technology Foundation. He is an alumnus of the Young Academy of the Royal Netherlands Academy of Arts and Sciences and an elected member of the IEEE Solid-State Circuits Society AdCom, the society's governing board.

Index

S.M. Kashmiri and K.A.A. Makinwa, *Electrothermal Frequency References in Standard CMOS*, Analog Circuits and Signal Processing, DOI 10.1007/978-1-4614-6473-0, © Springer Science+Business Media New York 2013

Printed in the United States
By Bookmasters